TOSCA
The Total Social Cost of Coal and Nuclear Power

TOSCA:

The Total Social Cost of Coal and Nuclear Power

Linda Gaines
R. Stephen Berry
Thomas Veach Long, II

Resource Analysis Group
Committee on Public Policy Studies
and Department of Chemistry
The University of Chicago

Ballinger Publishing Company • Cambridge, Massachusetts
A Subsidiary of Harper & Row, Publishers, Inc.

This book is printed on recycled paper.

Copyright © 1979 by Ballinger Publishing Company. All rights reserved. No part of this publication may be reproduced, stored in a retrieval system, or transmitted in any form or by any means, electronic mechanical photocopy, recording or otherwise, without the prior written consent of the publisher.

International Standard Book Number: 0-88410-086-3

Library of Congress Catalog Card Number: 78-26240

Printed in the United States of America

Library of Congress Cataloging in Publication Data

Main entry under title:

TOSCA, the total social cost of coal and nuclear power.

   Includes bibliographical references.
   1. Electric power-plants—United States—Costs.
2. Atomic power-plants—United States—Costs.
3. Electric power production—United States—Costs.
I. Gaines, Linda. II. Berry, R. Stephen, 1931–
III. Long, Thomas Veach.
TK1223.T17   338.4'3   78-26240
ISBN 0-88410-086-3

Contents

List of Figures · vii

List of Tables · ix

Preface · xi

Summary · 1

**Chapter 1
Overview** · 5

**Chapter 2
The Objective** · 9

**Chapter 3
Calculation of Costs** · 13
Method · 13
Contributing Costs under Normal Operation · 14
Costs Imputed to Accidents and Contingencies · 23
Concluding Remarks · 25

Chapter 4
Analysis 27

Description of Scenarios Considered 27
Costs Calculated from the Model 30
The Supply Problem 72
Extending the Calculations 76

Chapter 5
Concluding Viewpoints 87

Appendix 1
Cost Estimates 93

Direct Financial Costs 93
Indirect Costs 97
Safety and Damage to Human Health 99
Accidents and Contingencies 102

Appendix 2
The Program 105

Description of the Computer Program 107

References 117

Index 121

About the Authors 125

List of Figures

4−1	Consumer Demand vs. Time	29
4−2	Growth of Costs Through Time for Four Scenarios	32
4−3	Map of the Total Social Cost Surface with Construction Costs K_c and K_n as Variables, 4% Growth of Demand, 10% Discount Rate	33
4−4	Map of the Total Social Cost Surface with Construction Costs K_c and K_n as Variables, 7% Growth of Demand	37
4−5	Map of the Total Social Cost Surface with Construction Costs K_c and K_n as Variables, Discount Rate of Zero	39
4−6	Map of the Total Social Cost Surface with Construction Costs K_c and K_n as Variables and a Discount Rate of 20%	40
4−7	Map of the Total Social Cost Surface with Construction Costs K_c and K_n as Variables, Linear Growth of Demand	41
4−8	Map of the Total Social Cost Surface with Construction Costs K_c and K_n as Variables, No New Technology	42
4−9	Map of Total Social Costs with Human Cost H_c and H_n as Variables	44
4−10	Map of Total Social Costs with Capital Costs of Coal Plants K_c and of New Technology K_ν as Variables	47
4−11	Map of Total Social Cost Surface with Initial Fuel Costs as the Variables	48
A2−1	Flow Chart	106

List of Tables

3−1	Contributing Costs	15
3−2	Cost Estimates (1974$) for 1,000 Mwe Electric Generating Plants	16
4−1	Total Social Costs: Base Case, Zero Discount Rate	50
4−2	Total Social Costs: Base Case, 5% Discount Rate	51
4−3	Total Social Costs: Base Case, 10% Discount Rate	53
4−4	Total Social Costs: Base Case, 20% Discount Rate	55
4−5	Incremental Cost Factors for Category 1 Costs	58
4−6	Incremental Cost Factors for Category 2a Costs	58
4−7	Incremental Cost Factors for Category 2b Costs	59
4−8	Incremental Cost Factors for Category 3 Costs	59
4−9	Incremental Cost Factors for Changes in the Escalator β for Human Costs from Coal-Fired Plants	60
4−10	Total Social Costs: No New Technology, Zero Discount	61
4−11	Total Social Costs: 5% Discount Rate	62
4−12	Total Social Costs: 10% Discount Rate	64
4−13	Total Social Costs: 20% Discount Rate	66
4−14	Incremental Cost Factors for Category 1 Costs and No New Technology	68
4−15	Incremental Cost Factors for Category 2a Costs and No New Technology	68
4−16	Incremental Cost Factors for Category 2b Costs with No New Technology	69
4−17	Incremental Cost Factors for Category 3 Factors with No New Technology	69

4-18	Contributing Costs of Various Kinds	70
4-19	Electric Energy Generated According to Fuel	73
4-20	Amounts of U_3O_8 to Fuel the Nuclear Power System	74
4-21	1978 Increments to Capital Costs	80
4-22	1978 Increments to Operating Costs	80
4-23	1978 Increments to Fuel Costs	83
4-24	Sum of Increments to Total Costs, Based on 1974 to 1978 Changes in Direct Costs for the Commonwealth Edison Company	83
4-25	Total Costs, 1974 and 1978, Based on Direct Costs for the Commonwealth Edison Company	85
4-26	Breakdown by Category of Increments (1978-1974) of Costs for New Plants	86

Preface

The research on which this book is based was begun in 1975, soon after Linda Gaines came to The University of Chicago to work in the then informal group that became the Resource Analysis Group of The Committee on Public Policy Studies. The project was stimulated by at least two considerations, the intensity of the debate over coal and nuclear power, and the apparently irreconcilable figures then being quoted concerning the energy and materials requirements for nuclear power systems. The project began as an examination of the physical factors that go into power systems, but its scope expanded toward its present form as we gained perspective, self confidence and, we believe, the confidence of the people to whom we turned for primary data.

When we began, none of us was strongly committed rationally or emotionally to either nuclear or coal as a primary source of electric energy. And we still have no strong leanings; in our eyes, both sources have problems that cannot be dodged by trying to find cheap and expedient but inadequate ways around those problems. If we have any commitment at all, it is to reducing demand for electric energy by rational approaches to conservation and efficiency.

The outcome of our work is an analytic tool, a model, and a large set of numerical results of cost calculations. The analytic tool of Total Social Cost Analysis (TOSCA) is an approach that we believe can be applied at least as well to other policy problems as it has been used to compare coal and nuclear power. The model is intentionally simple because it is intended to be a readily available tool for a wide audience rather than an elaborate apparatus for the specialist. We wanted the model to be transparent enough to enable all its users to understand why each comparison comes out as it does. The world has need for both simple models and elaborate ones. When trying to decide what is important, one turns to the former; when fine-tuning an economy, the latter is appropriate.

The term "total social cost analysis" calls for comment. First, such an analysis must incorporate the uncertainties in costs and the implications of these uncertainties. Second, as the reader will discover, we believe there are considerations such as proliferation or disruption of the social structure that are so uncertain that it is meaningless to attribute a cost to them. Their uncertainties are as large as their potential costs; one might even question whether a particular choice of electric power system will lead to a net cost or a net benefit in such areas. Our attitude that some effects cannot be assigned meaningful costs might be taken by some readers to mean that "total social cost" is a grandiose misnomer for what we have tried to do. In defense of our use of the term, we can only say that those readers are free to include their own estimates of the costs of what we think of as political considerations—but we would insist that they estimate the uncertainties in their evaluations, and be prepared to come to a policy decision based not on a single estimate of a cost but on the range of uncertainties within which they think that cost may fall.

It is our belief that analyses such as ours should be indicative tools, not determinative forces, for social choice. We see nothing "wrong" with making decisions whose costs are not the lowest according to TOSCA or any other device for estimating costs. There may be sound social or political reasons for such choices, especially of the kind associated with great uncertainty and great risk. If we use our analysis in such a situation, its value lies in that it tells us the premium we pay to avoid that risk.

Throughout the work, we drew frequently on many people for advice and critical review of our progress. We would especially like to express our gratitude to those generous and helpful people who provided us with data or read and criticized the manuscript in its earlier drafts: Joseph G. Asbury, Norman Bradburn, Paul Chenea, Donald Collier, Gordon Corey, Gerald Decker, Richard Garwin, Werner Glass, Denis Hayes, Sandra Hebenstreit, John Holdren, Alvin Johnson, Franklin A. Long, Doan L. Phung, David J. Rosen, A. David Rossin, D.A. Vermilyea, Alvin M. Weinberg, Macauley Whiting, and A.M. Wolsky.

This work was initiated under a grant from the National Science Foundation, Office of Energy Research and Development Policy, and the work was concluded with the support of an unrestricted grant from the Exxon Company, USA.

TOSCA
The Total Social
Cost of Coal
and Nuclear Power

Summary

A method is given—and applied—to determine the optimum mix of fossil-fueled and nuclear-fueled electric power generating plants, for the United States, for the next thirty years. The criterion of judgment is the total social cost, including both the apparent or "internalized" costs and the hidden or "externalized" costs. The method involves making estimates of as many of the factors as possible that contribute to the costs, finding the total cost implied by these estimates, and then varying the estimates widely to determine how sensitive the choice of mix is to the estimates. This method of *total social cost analysis* is, for convenience, called TOSCA. The choice of name is meant to imply the aim, but not necessarily the actual achievement, of the approach.

The purpose of this book is twofold. One is to provide a method simple and general enough to be used by anyone interested in the nuclear power question. One can see how his or her own estimates of the costs of individual factors change the total social costs and shift the mix of nuclear and fossil plants that has the lowest total social cost to the society.

The other purpose is to use the method to provide a set of working estimates of the total costs of alternative mixes with realistic costs appropriate to various regions and guesses about what the future costs will be. The results of TOSCA show that the optimum mix is sensitive to some factors, such as region, future fuel costs, and the rate of growth of demand for electricity. It is insensitive to other factors, such as the cost of safeguards against sabotage of power plants and the diversion of nuclear fuel. For regions such as the Chicago metropolitan area or the New York–Boston area, our estimates of the lowest cost mix are heavily (50 percent or more) or entirely nuclear; for areas such as Wyoming, the optimum mix is predominantly or totally coal-fueled. If capital costs are taken to be similar to those in the Chicago area, whether or not new technology is developed to replace nuclear capacity by the end of the thirty-year period, and if the cost of fossil fuel goes up at a rate comparable to or faster than the cost of nuclear fuel (measured on the basis of equivalent amounts of generated electric energy), the model then indicates that the optimum mix is all nuclear-fueled electricity generation. If the price of uranium goes up faster than the price of coal, the optimum mix is a combination, whose relative proportions of nuclear and fossil plants depend on the shape of the growth of demand. If demand grows at a high rate, for example, 7 percent per year, the best mix is heavily weighted with fossil plants. Lower demand over time tends to favor a higher proportion of nuclear power in most scenarios.

One may ask whether TOSCA and its results could be used by someone making a policy decision at a national level. The answer is "yes," but not in the simplistic way of opting to encourage the exclusive use of coal or nuclear power everywhere in the United States at the expense of the other. The conclusions of TOSCA are that the lowest cost choice varies from one location to another and that an intelligent national policy would be sensitive to these variations.

There will always be considerations to which no meaningful cost can be assigned, which are nevertheless important to our choices. By generating tables of total costs for the full range of options, from all coal to all nuclear power, we tell ourselves the extra price we should be prepared to pay if, for reasons of security, for example, we choose a mix whose costs are not the lowest. In this way, TOSCA should be used as an indicative tool for policy determination, and not as a determinative device. The factors that were varied in this study were as follows:

1. Demand, as a function of time
2. Initial capital costs, including carrying charges
3. Operating and maintenance costs, apart from fuels
4. Fuel costs
5. Research and development, including advanced system costs (discussed below)
6. Property damage from pollution
7. Heat dissipation

8. Health effects of pollution
9. Safety of normal operations
10. Spent fuel storage
11. Large accidents
12. Terrorism, sabotage, and diversion
13. Discount rate of all factors except human life
14. Discount rate of human life

The assumption is made that generating capacity is the minimum capacity that meets demand. The capacity considered here should be thought of as base load capacity because this is the normal role for new facilities.

Factors 2 to 12 fall into three principal categories, with minor variations, according to how they depend on the construction of new and replacement capacity, and they specifically depend on:

ΔN, the new increment of capacity added in each year

N, the total capacity in operation during the current year

η, the total energy generated from the beginning of the period up to the current year

t, the time from the beginning of the period

Initial capital investment depends only on the new increment. Operating and maintenance costs of each mode of generation depend only on the total capacity of that mode. Lifetime average costs are used for each mode.

The model calculations include the possible kinds of behavior for fuel costs, that is, costs simply that increase with time only or fuel costs that increase with total energy generated to date (to reflect the rising cost of a depletable resource). Although we have not done so, it would have been possible to model decreasing real costs of fuel. The model for research costs for coal and nuclear technology depends on total capacity in operation; the cost of research for new technology is taken proportional only to time. The cost of spent nuclear fuel storage is a function of the total fuel used and therefore of the total energy generated. Costs of property damage and air and water pollution control and safeguards depend on capacity in use but may have some dependence on total energy generated if the impact of spent fuel is considered.

The results of the calculation are presented in two ways. A set of figures, like maps, are drawn with all but two of the costs fixed at a "base case" or at another case of interest. The two axes correspond to two costs whose relative influences are to be compared, such as the initial capital costs of fossil and nuclear plants. Curves and lines of constant total cost are constructed and the space in each map is divided into regions corresponding to different mixes, that is, all nuclear, all fossil, or some of each fuel, for situations in which these are the only options.

The other calculation is a set of tables of total social costs for the thirty-year period for a large assembly of choices of discount rate, the growth of demand, new technology, and fuel costs. These costs are all computed for scenarios that are all coal, 75 percent coal, a 50–50 mix, 75 percent nuclear, and all nuclear. This way, one can identify not only the lowest cost mix, but one can also see how sensitive the total social cost is to the choice of mix. These tables are evaluated from a single base case for which capital and operating costs are taken to correspond to the Chicago–Commonwealth Edison situation.

For example, if the discount rate is 10 percent, the growth of demand for electricity 4 percent per year, the costs of coal and uranium maintain the same relative price, and the direct costs taken to be the 1974 costs for the Commonwealth Edison Co., then the optimum mix is all nuclear. This attains whether or not new technology is introduced and whether or not human costs are discounted. The present costs for the thirty-year period in these sample all-nuclear scenarios are $284 billion (no new technology and discounting human costs at 10 percent per year); $382 billion (no new technology and zero discount rate for human costs); $563 billion (new technology and human costs discounted at 10 percent per year), and $648 billion (new technology and no discounting of human costs). Suppose, on the other hand, that the cost of uranium escalates, doubling in twenty years, then the total cost of an all-nuclear scenario is the same as that of a half-coal, half-nuclear choice if new technology is introduced ($600 billion, with human costs discounted). With no new technology, the half-coal, half-nuclear choice is $598 billion, $8 billion less than the all-nuclear option.

A set of supplementary tables permits one to calculate the corresponding total costs of the five representative fossil-nuclear mixes with an arbitrary set of input costs. The model was constructed so that these calculations can be done in minutes with pencil and paper if the alternative costs are known or assumed.

Chapter 1

Overview

Of all the issues concerning energy or resources, the balance of nuclear- and coal-generated power remains the most fiercely debated and most emotional subject. Unfortunately, the heat of the argument cannot generate electricity, and it often seems unable to help resolve the question of what we should do. In recent months, some progress has been made toward calm analysis of aspects of the problem. Manne, for one, has examined the direct costs and substitutions among fuels, with feedback to show the effect of price changes on demand.[1] Earlier models had assumed that demand is exogenous.[2-5] The Council on Economic Priorities[6] has examined costs of alternative options with some attempt to include estimates of indirect costs as well as direct costs. The Nuclear Energy Policy Study Group report[7] (the "Ford-Mitre study") has pursued a total cost approach similar in goals and complementary in method to the one presented here.

The central question, particularly for committed opponents of either source of power, is the importance—call it the costs—of the indirect and externalized effects of using coal or nuclear energy to generate electricity instead of employing alternatives. To many of the enthusiastic supporters of nuclear power, for example, the indirect costs appear less important than the uncertainties associated with direct costs, uncertainties over the cost of capital, the cost of fuel, and the cost of meeting regulatory constraints. The same holds for the proponents of coal-fired power generation. Until now, most examinations of the alternative midterm sources of energy have dealt in depth with one or perhaps a few of the costs. In this book, we attempt to confront the problem in as many aspects as we can in hopes of moving the debate to a new level. The Ford-Mitre study is more concerned with the quality of each cost estimate, while we are more concerned with the relations between the costs and the choice of a mix of energy sources.

The goal of this study has been to develop tools to answer the following questions:

1. Above all, what is the optimum mix of fossil- and nuclear-powered electric generating plants for the United States for the next thirty years, when we base this choice on minimizing the total social cost as best we can by estimating as many indirect and externalized costs as possible?
2. How sensitive is our decision to the estimates of particular costs and other quantities that influence the total cost?
3. How much can we expect to pay if we choose an option different from that with the minimum "total social cost"?

The method we use is an analysis of total social cost, which we abbreviate with the acronym TOSCA.

This approach has obvious roots in the approach of Pigou, who built a theory of public expenditure on the idea that "public" choices—choices made by governments—should provide maximum social benefits for the minimum total social cost.[8] The aspect of Pigou's thinking most relevant to our analysis is his belief that it is possible to achieve meaningful results in attempting to estimate social costs of all sorts. Pigou argued that a proper evaluation of alternatives required a quantitative estimation of their social benefits as well. Here is one point at which our problem is simpler than that of Pigou. Our question is deliberately framed so that we only ask about costs and not about benefits because all the alternatives deliver the same benefits.

We deal with a predetermined consumer good to be delivered, the electric power necessary to satisfy an exogeneous demand; our problem is only one of finding the best mix of means to supply that good, based on total social costs of the alternatives. Pigou attacked the larger and much more intractable problem of selecting what goods to supply from the public expenditures. This requires estimating total social benefits as well as total social costs, a far more difficult and ill-defined problem.

There is another difference—we are examining a public choice, but not a governmental expenditure. The selection of sources of electricity is made in a context in which decisions have traditionally been in the hands of private regulated monopolies with oversight, review, and veto power in the hands of the public agencies that regulate them.

By presenting specific demand scenarios, we do not mean to espouse any one of them as optimal. No one has argued more fervently than we that energy demand is elastic and sensitive to substitutions. Our four demand scenarios are chosen to indicate the influence of the wishes of consumers on the optimal mix of coal-fueled and nuclear power.

We have tried to assemble all the costs, internalized or not, that might be considered important; to make or find "most reasonable" estimates of those costs, and also estimates of upper and lower bounds, whenever appropriate, and to carry out analyses of total costs to determine the sensitivity of the choice to the individual costs. We believe our estimates and ranges are realistic. However, it is far more important, in our view, for the nuclear-or-fossil debate to be put as a sensitivity study of total costs than for our particular values to be precisely correct or for the conclusions presented here to be accepted universally. It is our hope that this book will help to clarify which factors are important to study in more detail, which factors can be removed from the argument because even the upper limits of their costs are small, and which factors are undeniably important but uncertain. By taking a broad view, we hope to help separate questions that are primarily technical from questions that are essentially social or political choices. We are quite prepared to see controversy regarding improvement in our estimated values and, doubtless, refinement of the model as well. But we want very much to see the debate remain at the level of the evaluation of total social costs rather than return to its formative phrasing in terms of fragmented, unconnected questions.

In one sense, this book is more concerned with a method than with an answer. We present a tool from which anyone can easily compute his or her own estimates of the *total social cost* associated with an arbitrary mix of two technological alternatives, estimate the sensitivity of the outcome to the estimates, find the proportion of the two technologies that has the lowest total social cost and know how firm that answer is likely to be. Even in our calculations, we find different optimal mixes for different regions of the United States. One should not be troubled that the method may give answers that are very sensitive to the input; such findings are extremely useful because they indicate which costs must be estimated accurately if the answers are to be convincing.

We use the near acronym TOSCA for TOtal Social Cost Analysis to denote the mapping of total social costs and the estimation of their sensitivity to estimated individual costs when one takes into account every factor one can think about. The TOSCA procedure is straightforward; determining meaningful input data is sometimes very difficult as we shall see later.

Before embarking on a discussion of our procedure, we must provide several disclaimers. First, the approach does not address conservation explicitly. However nothing in the model is inconsistent with energy conservation, and some of the demand options are the sort that would follow from a husbandry effort. Second, we assume demand is exogenous, but we attempt to provide approximately for the kind of feedback given by price-sensitive models (such as Manne's) by considering a variety of demand curves. Third, the problem as phrased here deals specifically with the U.S. domestic electricity supply, and while domestic problems of diversion, protection, and accident are included, problems of proliferation of nuclear weapons are not considered. Fourth, our analysis is not long term; the thirty-year model does include provision for development of alternatives to supply electric power at the end of the thirty-year period, but we have made no attempt to compare advanced technologies of breeder reactors, solar energy, nuclear fusion, or other more exotic possibilities. Hence the options in our model are simply fossil-fueled and conventional nuclear "burning" plants—BWRs and PWRs (boiling-water reactors and pressurized-water reactors). New technologies are incorporated as an alternative for the years after 1990 in a manner that leaves most of their detailed characteristics unspecified.

Chapter 2

The Objective

The United States faces a sequence of near-term decisions about how to supply energy. Many of the most costly choices concern the supplying of electricity. Inevitably, we must decide what fuels to use to generate electricity, that is the least substitutable of all the forms of energy that we commonly use. In thirty or forty years, there will probably (and hopefully) be other alternatives, but between now and thirty years hence, we must plan on the basis that there are only two alternatives for fueling the overwhelming majority of our electric generators, particularly for new capacity—coal and uranium. Moreover the uranium can only be used in conventional "burning" reactors, either LWRs (light water reactors, either pressurized—PWRs—or boiling—BWRs) or HWRs (heavy water reactors) such as the Canadian deuterium (CANDU) system.*

The immediate object of this book is to develop, from a set of estimates of all the significant costs and factors directly affecting the costs of coal-fired generation and nuclear generation of electricity, the composition of the mix of fossil and nuclear plants that minimize the total social cost. In short, we would like to be able to say how many fossil plants and how many nuclear plants we should build during the next thirty years. And secondarily, we would like to know how much a change in the estimate of each input datum will affect the composition of that optimal mix.

The cost and related variable factors fall into several categories. Obviously the direct costs, both of capital equipment and operation,

*At present, the United States is not using CANDUs, and they are not included in our analysis.

are large and important. Fuel costs are probably the most uncertain of all the costs of the system and among those to which the choice is most sensitive. The discount rate must be considered as an important variable input datum. The curve of demand through time, while not a cost, is an important variable in our approach because we assume that electricity production meets demand with as little overcapacity as possible. We simulate feedback, rather than introduce it explicitly, by using demand curves of various shapes.

Indirect costs in our model include costs of research and development, because we assume that the supply system will have to continue supplying electricity smoothly after our thirty-year interval ends; costs of property damage; of human health and safety; of spent fuel storage; of potential accidents; of safeguards against sabotage and terrorism; and of global and regional climate changes due to discharge of waste heat and carbon dioxide.

Rather than estimate benefits explicitly, we formulate the problem in terms of the costs of alternatives. The results are equivalent if we have truly included all the significant costs.

Certain kinds of cost have not been included explicitly. For example, the cost of water needed to supply fuel is not included except insofar as it is included in the cost of the fuel. This procedure is correct provided that the cost of fuel reflects reasonably accurately the total cost of the water. If the externalized costs of the water are small relative to its purchase cost, the price of coal or uranium will incorporate most of the cost of water. If the externalized costs of using water were large, one would want to augment the list of costs to include them.

The most obvious omission from this book is the issue of proliferation of nuclear weapons. While this book confines itself to the selection of electrical generating systems in the United States, the argument is sometimes made that U.S. decisions regarding nuclear power have far-reaching implications elsewhere. The danger perceived is that decisions in the United States to adopt nuclear power will increase the likelihood that other nations will also do so and that nuclear facilities ostensibly devoted to electric power generation can be diverted to manufacturing material for nuclear weapons.

This is an important social issue, but it is an issue for which an attempt at quantification of the costs would be delusive. The probabilities depend so sensitively on matters specific to individuals and political situations of the moment that proliferation should be treated as a political rather than an economic problem.

Similar to proliferation in this regard is the question of social effects of extensive safeguard measures. Insofar as safeguard measures would require restrictions on personal freedom, most people would say they would entail some social cost. But estimating the cost of the impacts on personal freedom is an issue that transcends simple economics; it, like proliferation, is a philosophical and political question for society.

Many of our readers will doubtless have costs in their own minds that we have neglected. Our model is intended to provide them with a simple device with which they can compute the effect of their cost estimates on the total cost and, then, on the composition of the optimum mix. It is only necessary to make an explicit assumption about how the new costs depend on time, fuels, and the number of plants of each kind. We later illustrate how to use the model, in the hope that the procedures will be as transparent as we can make them.

Chapter 3

Calculation of Costs

This chapter describes the way each cost is computed. Here we look at how each is likely to depend on the number and kind of plants built of each type, on the fuels consumed, and on time. Discussion of the numerical estimates that we chose for a base case and for the ranges of the costs can be found in Appendix 1.

METHOD

We identify twelve categories of costs and differentiate between discount rates for human life and for all other factors. One must make specific assumptions about how each of these cost factors depends on the mix of fossil and nuclear plants, on the total electricity demand, and on time. One estimates the magnitude of each cost to see whether any is small enough to be neglected for the interval of this book. Then one determines a more reliable estimate of each cost and estimates its likely range of uncertainty. Some of the cost factors must somehow be converted from nonmonetary costs into monetary terms, specifically in $/unit. The others appear naturally in monetary terms of $/unit. At this point, one has ranges of estimated alternative costs of the independent input variables. To calculate a total social cost from these figures, one must have a scenario. In our approach, the selection of discount rates and a demand curve define the scenario completely because we assume that power production meets the demand. (This is the reason for including demand functions that simulate behavior such as price sensitivity.)

The first step in calculating the total social cost is in the identification of those costs that make significant contributions. They must include internalized and externalized factors, that is, the utility company's costs plus costs to the public in the form of pollution and safety hazards and costs to the government such as research expenditures. One lists the possible costs and estimates the order of magnitude of each.

At this stage, negligible items are dropped from further consideration. The dependence of each kind of cost on the number of plants in service, time, fuel used, and so on, must be determined. The contribution in year t of each kind of power generation to the total cost can then be calculated for each scenario. All costs are converted to a common unit (\$/year in this case; see discussion later of the valuation of human life) and summed over the n kinds of power plant—coal, nuclear, and any new technology—to obtain the cost at time t, $\Sigma C(n,t)$. The costs are then discounted at a rate r and summed over time to give the present value of the total social cost of each alternate plan for supplying electricity.

$$\text{Total social cost}, S.C. = \sum_{\substack{\text{years } t, \\ \text{costs } j}} (j^{\text{th}} \text{ cost in year } t)(1+r)^{-t} \quad (3-1)$$

Equation 3-1 is the basic equation from which all the hypothetical futures are evaluated in our model.

It should be noted that this total can be expected cost, where the cost of each possible event (such as an accident) is weighted by its probability of occurrence, or the perceived cost, where certain events may be weighted more heavily because they are considered morally or politically unacceptable (risk aversion). The expected cost is simply the sum over possible events of the probability of each times its cost.

CONTRIBUTING COSTS UNDER NORMAL OPERATION

The contributing costs are described here and summarized in Table 3-1. Present values for these costs are listed in Table 3-2.

Direct Financial Costs

The three terms included here are initial capital investment, K, operation and maintenance costs, L, and fuel costs, E. Distribution costs are excluded since they are the same for all plant types and

Table 3-1. Contributing Costs.

Contributing Cost	Type	Comments	Who Pays	Dependence	Category
1. Direct costs	Initial capital investment	Include carrying charge	Utility	$K\Delta N(t)$	1
	Operation and maintenance	Includes insurance, taxes	Utility	$LN(t)$	2a
	Fuel	Includes coal waste (ash) disposal, and transportation	Utility	$EN(t)(1+\beta\eta(t))$	2b or 4
2. Indirect financial costs	Research		Government, manufacturers and indirectly utility	$RN(t)$ or $(R^0 + R^1(t))N(t)$	2a or 4
	Property damage	Acid rain, soot	Public	$N(t)$	2b
	Heat loss	Eutrophication of lakes, ice cap	Public	$TN(t)$	2a
3. Human health-safety	Pollution and operating safety	Split into 2 categories, safety and safeguards	Public, government	$HN(1+\beta\eta(t))$	2b
	Spent fuel storage	Put with direct financial costs, related to safeguards	Utility; public if mismanaged	$S_0\eta(t)$ or $S_0^1 N(t)$	2a or 3
4. Accidents and contingencies	Large accidents	Insurance equivalent	Insurer (government or private)	$AN(t)F$	2a
	Terrorism and sabotage	Safeguards probably cheaper	Utility; public if inadequate safeguards	$(TS)N(t) + (TS')\Delta N(t)$	2 or 3

Table 3-2. Cost Estimates (1974$) for 1,000 MWe Electric Generating Plants *(Ranges Given in Parentheses)*

		Type of Plants				
		Coal				
Constant	Units	With Scrubbers	Without Scrubbers (low-sulfur)	LWR	LMFBR	Solar
K	$ × 10^6$ charged in 1st operating year	420 (162–935)	295 (175–565)	421 (182–863)	600 (500–1000)	1000 (750–2000)
K'	$ × 10^6$ charged in year of decommissioning			24 (3–45)		
L	$ × 10^6$ plant year (py)	17.1 (−3–17.6)	11.4 (9.3–11.4)	11.4 (6.2–11.4)	12.5	40
E	$ × 10^6$/py	52.6 (≅9.4–52.6)	84.2	28 (15.3–30.5)*	20	0
R^0	$ × 10^6$/y or py	1	1	5	5	2
R_v^0	$ × 10^6$	—	—	—	500	15
R_v'	$ × 10^6$/y	—	—	—	80	50
PD	$ × 10^6$/py	3 (0.8–30)	3	0	0	0
T	$ × 10^6$/py	∼0	∼0	∼0	∼0	∼0
H	$ × 10^6$/py	16 (2–100)	8 (0–12)	1 (0.01–1)	3	2
S_o	$ × 10^6$/py (1×)	0	0	2.5 (0.5–17.5)		0
S_o'	$ × 10^6$/py^2 (perpetual)			0.3		
A	$ × 10^6$/py			0.05		
TS				0.5		
β_0		$2 × 10^{-5}$ (0–0.0003)	$2 × 10^{-5}$ (0.0003)	$2 × 10^{-4}$ (0–0.0003)		

therefore do not change the optimum mix. All are paid by the utility company. Initial capital investment is the financial cost incurred for construction of a new generating facility, including carrying charges on the capital and property taxes as well as the cost of the land, any cooling towers or ponds, and the initial fuel assembly. Expenditures are proportional to the number of new plants of each type constructed and assessed in the year the units go into service.

$$K(t) = K_c \Delta N_c(t) + K_n \Delta N_n(t) + K_\nu \Delta N_\nu(t) \qquad (3-2)$$

Each $K(t)$ has the form of a unit cost, K_c^0, K_n^0 or K_ν^0 times $\Delta N_i(t)$, the number of new plants brought on line in year t. We employ a one-time cost for construction and carrying charges in order to examine the sensitivity of total cost to costs proportional to the number of new plants added. (Subscripts "c" = coal, "n" = nuclear, ν = new.)

The cost of decommissioning plants must also be added. It is proportional to the number of plants built thirty-five years earlier (see the first section in Chapter 4 for discussion of plant lifetimes). This is $\Sigma_i K_i' \Delta N_i(t-35)$. Thus the decommissioning cost for a plant built in 1960 would appear as a one-time cost in 1995, and the decommissioning cost of a plant built in 1980 would not appear until 2015. Discounting makes the first of these small, and the second would not appear at all in our figures.

An alternative to the one-time cost assessment is to amortize construction costs and set aside funds for decommissioning throughout the life of the plant. With this choice, the construction and decommissioning costs enter in the calculations exactly as operating costs do.

Operation and maintenance costs consist largely of wages but also include those of routine repairs and other overhead expenses of the utility. In our model, this expense is assumed constant over a plant's lifetime, and the annual total is simply proportional to the numbers of each plant type in service in that year, $N_i(t)$.

$$L(t) = L_c N_c(t) + L_n N_n(t) + L_\nu N_\nu(t) \qquad (3-3)$$

where each L is of the form, for example, for coal plants,

L_c = (annual unit plant cost of operating a coal plant)

and so forth, for coal, nuclear, and new technology plants, respectively. The fuel cost is the expenditure by the utility for fuel to operate its generating plants, including transportation as well as interest on any fuel inventory. The expense is proportional to the number of plants of each type in service and their capacity factors.* We have made the simplifying assumption that the capacity factor (cf) is constant over a plant's lifetime. (This does not affect results if a large number of plants are in service.)

It is extremely difficult to predict future fuel prices, but it is commonly believed that they are more likely to rise than drop in constant dollars. In our treatment, we use three different models to represent the evolution of fuel prices. In the first model, we assume a price proportional to the quantity of that fuel already used. This model is supposed to represent the introduction of more expensive, lower grade ores and higher exploration costs as cheaper or more accessible ores are depleted.** Formally, the fuel cost can be written as

$$E(t) = \Sigma_i E_i \, (cf)_i \, N_i(t) \, [1 + \beta_i \, \eta_i(t)] \qquad (3-4)$$

Here E_i is the annual fuel cost for the ith kind of plant operating in year 0 at 100% capacity. Also, $\eta_i(t)$ is the integrated total power generated by the ith type of plant to the year in question (i.e., total energy), and β_i is the price rise parameter.

In the second model, we consider scenarios in which fuel prices rise as simple functions of time with the prices of other factors. The third model assumes that the fuel prices are constant.

Costs of nuclear fuel were estimated on the assumption that wastes would be stored or discarded but not reprocessed. This is in accord with present practice. If reprocessing becomes a feasible option, the associated adjustments could be incorporated in the costs of fuel and waste disposal.

Indirect Financial Costs

Research. Large sums of public and private money are spent for research on alternate methods of electricity generation. This expense may not be of direct concern to the utility company deciding what type of plants to build, but it must enter into a calculation of total cost. Money is spent both to develop new technologies for future electricity supplies and to improve commercial technologies (coal-cleaning and nuclear safety, for example.)

---

*Defined as actual average power output/rated capacity.

**For coal, the supply of easily available fuel is not immediately limited, but the supply of miners may be; even rising labor costs could conform roughly to this model.

Research on a new technology is assumed to start from present levels and increase linearly to a maximum. Its cost has the form

$$R_\nu(t) = R_\nu^0 + R_\nu^1 t \quad \text{until the cost reaches} \qquad (3-5)$$

$$R_\nu^{\max} \quad \text{and constant thereafter}$$

until commercialization or failure. The calculations include both scenarios in which research succeeds and new technology is brought into use and scenarios in which the research fails and the system reverts to building its current mix. We also consider the case where research on new sources is abandoned (research cost = 0) and all power is supplied from presently commercial plant types. In this situation,

$$R_\nu(t) = 0 \quad \text{for new technology} \qquad (3-6)$$

For each plant in commercial service, a fixed annual research cost is assessed:

$$R_n(t) = R_n^0 N_n(t) \quad \text{for improving nuclear plants and}$$

$$\qquad\qquad\qquad\qquad\qquad\qquad\qquad\qquad\qquad (3-7)$$

$$R_c(t) = R_c^0 N_c(t) \quad \text{for improving coal-fired plants}$$

This cost is added to operating and maintenance expenses.

Property Damage. Pollution from the burning of fossil fuels soils buildings, causes corrosion, and damages crops. These costs are external to the utility, but they must be incorporated in the total social cost.

Costs related to mining (either coal or uranium) apart from reclamation, such as costs of land subsidence and acid drainage damage, are unknown and are not included. When land is reclaimed in a midwestern strip mine, the reclamation costs are estimated to be less than 5 percent of the fuel costs for mining[9] and fall within the uncertainties of the total fuel costs. Reclamation costs may be significantly higher for other locations. We assume that property damage costs are directly proportional to the quantity of fuel used per year and therefore add an annual increment to the cost of fuel.

Heat Loss. For every kilowatt-hour (kwh) of electricity produced, approximately 2.4 kwh of heat is eventually released into either the water or the air. An increase in the water temperature leads to greater biological activity[6] ($\sim 5\%/°F$) in the body of water. The costs from this can be negative (i.e., there can be benefits) if fishing yields are increased or positive if algal growth speeds eutrophication of a lake. (Damage to the lake might be avoided and profits made if the algae were harvested for food or fuel.) The net cost must be evaluated in the detailed context of each generating facility. We have not tried to estimate costs associated with heat put into lakes, rivers, and oceans.

Increases in ambient air temperature are significant locally for present levels of energy use. The effects of the urban "heat island," such as increased cloudiness and temperature, are well known.[10,11,12] We shall later estimate that portion of the cost due to power plant waste heat. In cold areas an increase in average air temperature would save money spent for heating fuel, while in warmer areas the increase in air-conditioning uses more fuel than is saved by the decrease in winter heating. This thermal effect is proportional to the power generated each year (number of plants in service) for small changes ($< 1°F$) but may become nonlinear as feedback effects come into play.

$$T(t) = T_c N_c(t) + T_n N_n(t) + T_\nu N_\nu(t) \quad \text{for small changes}$$

(3-8)

For very high levels of power use, global climatic effects are possible, conceivably even melting the Greenland ice sheet. Meteorologists have made order of magnitude estimates of critical temperature changes:

> Budyko (1970) and Sellers (1970) argued that it would take only a change of 1.6 percent in the solar radiation available to the earth to lead to an unstable condition in which the snow cover could advance to cover the continents all the way to the Equator; and in the process of doing this the albedo would be raised by the greater snow cover to the point where the oceans would eventually freeze.... This illustrates the delicacy of the thermal balances of our planet.[12a]

Calculations have shown that a 1 percent change in the solar constant (or equivalent anthropogenic heat) would cause between 1 degree and 5 degree Centigrade change in the mean surface temperature of the earth.[12b] A group at the National Center for Atmospheric Research[12c] performed a computer experiment that showed that a 5 degree Centigrade rise would be expected from a thermal activity level 100 times larger than the present level. Thus we can consider

100 times present energy use to be a global upper bound. With the largest growth rate we considered, 7 percent per year, power levels rise only by a factor of six by the year 2000, so no global heat effects were added to our total cost.

We have not ignored the possibility that atmospheric circulation patterns and heat balance may be unstable enough to be modified seriously by small perturbations. An increase in the rate of release of carbon dioxide or heat could change these patterns in a way that would affect the productivity of, for example, the midwest grain belt. It is altogether conceivable that costs from such climate modification would be so large as to make the remainder of this analysis irrelevant. Until these phenomena are better understood, their probabilities cannot be estimated with enough confidence to be the basis of useful estimated probable costs.

Human Health and Safety Costs

The cost to human health from the operation of a power plant and its support facilities (mines, railroads, etc.) is generally cited in numbers of lives or man-days lost per year. In the present context, we must consider the effects of chronic exposures to hazards and the possible accidents, small and large, that could occur with fossil or nuclear power generation. The cost of lives lost by accidents can be incorporated in our total costs by estimating the cost of a single life, the probability of each kind of accident, and the probable number of lives lost. Effects of chronic exposures are taken as certain and therefore need no additional factor to reflect their probability of occurrence. The valuation of a single life has been discussed by several writers.

Three criteria that lend themselves to quantification are willingness to pay (e.g., for insurance), discounted future earnings, and payments for risk either taken or avoided.[13-16] Mishan,[13] using willingness to pay, discusses a theoretical cost–benefit analysis of a single death including financial aspects and direct and indirect risks (concerning oneself and others.) Schelling[14] considers the losses of income, taxes, and charitable contributions, as well as the amount of money a person would pay to avoid a perceptible probability of his or her own death. He states that inquiry among professional people suggests a random life among them would be worth between 10 and 100 times a year's income, or on the order of a million dollars. Federal requirements for industrial safety require $1 million or more per expected life saved, but Federal Highway Administration spending is lower, that is, between $35,000 and $1 million per expected life saved.[14,15] In our model we use the value of $1 million for any single life and assume it is independent of the number of lives saved or lost. A decisionmaker using the model could easily modify this estimate to study the robustness of the conclusions to this assumption; for example, Cohen has taken $10 million for his calculations.[17]

There is a rational basis for attributing a higher cost to aggregated deaths than to isolated deaths because the former could entail special costs of economic disruption, social support, and welfare assistance. Such increments could be incorporated into TOSCA directly or as a risk aversion function.

Pollution and Operating Safety. Mining injuries and diseases (from coal dust, radon gas, or other mining residues) as well as fuel transportation accidents are proportional to the quantitues of fuel handled. Harmful emissions such as SO_2 and particulates from fossil plants or radioactive materials from nuclear reactors or fossil plants are assumed directly proportional to the quantities of fuel burned. These health effects are thus proportional to the quantiy of electricity generated by each plant, which is the product of its installed capacity and capacity factor. In addition, some of these effects may be aggravated as more and more fuel has been burned and pollutant concentrations in the atmosphere (or water) increase. For instance, it has not yet been established whether damages to human health are proportional to SO_2 levels, increase faster, or exhibit some sort of threshold effect[18] when the concentration of pollutant is low. We assume fossil fuel health hazards may be cumulative and in that case would rise linearly with the amount of fuel used:

$$H(t) = \sum_i H_i \, (cf)_i \, N_i(t) \, [1 + \beta_i \, \eta_i(t)] \qquad (3-9)$$

If technological change or other scale effects are believed to modify the severity of these damages over time, one can assume a higher or lower value of β_i to adjust the model. Incidentally, this functional form is the same as that for the variation of fuel costs with time.

One could ask what the total costs would be for a coal-burning system so safe and clean that its costs in human lives and health are negligible. Unfortunately, we have no basis for estimating the costs of technologies that could achieve this goal, and thus we do not pursue this calculation.

Spent Fuel Storage. Spent fuel storage costs may appear as direct financial costs, but they are included here because mismanagement of spent fuels presents a potential danger to human health. Two possible forms for this cost are examined. The cost may be proportional to the total amount of fuel used to date if the residues remain as a hazard and must be monitored for leaks and protected from sabotage in perpetuity:

$$S(t) = \sum_i S_o \eta_i(t) \qquad (3-10)$$

Here, S_o is the cost of storing the waste from one unit of exhausted nuclear fuel. However, if the wastes could be recycled or permanently disposed of with a one-time cost, the cost would be proportional to the quantity of fuel used in each year:

$$S(t) = \sum_i S_o^1 N_i(t) \qquad (3-11)$$

The unit cost factor S_o^1 in Equation 3-11 is of course not the same as that for extended care, S_o of Equation 3-10. The net cost could then reflect any value attributed to the recycled fuel. The most elaborate treatment of this topic is that by Cohen.[17]

COSTS IMPUTED TO ACCIDENTS AND CONTINGENCIES

Thus far, we have considered only the cost of "normal" plant operation.* We now discuss abnormal events of very low probability and high cost. The cost of losing n lives in an accident from the ith kind of power plant, $A_i(n)$, is large. Its probability $p_i(n)$ must be small. If we can even think of building the plants, the expected cost of accidents, $[A_i(n) \times p_i(n)]$, is very small. However, the perceived cost, $A_i(n) \times p_i(n) \times f_i(n)$, may not be small because of risk aversion. This aversion is represented here by the factor $f_i(n)$, which may be much larger than one. No matter how small $p_i(n)$ is the event is possible, and even the extremely remote possibility of a large cost may be viewed as unacceptable for moral, political, or financial reasons.

We have considered costs associated with two situations peculiar to nuclear facilities (LWRs, breeders, and support facilities), that is, the costs of large accidents and the costs of security against terrorist acts. We have not tried to evaluate the costs of different levels of accident avoidance; this certainly would be an appropriate refinement of the present analysis. We have also not tried to estimate the probable social cost of a terrorist act, that is, the cost of the failure

---

*Recall that these included commonly occurring mining and transportation accidents.

of a supposedly adequate security system. This could be done by adding to the security cost the estimated cost of a terrorist act, multiplied by its probability, which presumably declines as the direct cost of the guard system goes up.

Accidents

Accidents are not unique to nuclear power. The high probability of small accidents is tolerated and treated as "normal operation" for coal plants. The costs of "routine" accidents are already included under Pollution and Operating Safety. If we were to include new oil-fired reactors in our fossil-fueled mix, we would be faced with the possibility of oil tanker leaks, breakups, and explosions. For the few new hydroelectric generating stations that will be built, the consequences of dam breakage should, in principle, be considered and so should the accidents accompanying support technologies such as those for disposal of nuclear waste.

What is unique to nuclear power is the combination of low probability and large potential damage of such major events. Oil tankers do founder, and dams do break. Because no explosive or other major nuclear accident has ever occurred in a power plant, it is difficult to evaluate either the consequences or the public response. It may be that the public would be as indifferent to a few deaths occurring annually from nuclear plants as it is to deaths from coal mining and burning and yet find unacceptable the possibility of a large number of deaths even if the probability is small. This possibility is probably greater for nuclear plants because the number of potential lost lives is so much greater than for coal plants. We write the perceived direct cost of accidents:

$$A(t) = \sum_i N_i(t) \sum_n f(n) A_i(n) p_i(n) \qquad (3\text{-}12)$$

The total cost is unaffected if a change in the expected accident cost estimate A_i is accompanied by an equivalent change in the opposite direction in the risk aversion factor.

There are two extreme responses to accidents; all real cases will lie between them. One extreme is acceptance, as plane crashes or dam breaks are accepted. Here the risk aversion is roughly unity. A small sum of money is spent in investigations and attempts to prevent any similar malfunctions, but the social cost is essentially the direct cost. At the other extreme is uncompromising resistance, which might occur as a moratorium on both operation and construction of nuclear facilities. This could include closing down all light water reactors and breeders as soon as possible (about five years) and replacing them with other facilities, presumably coal plants. This case corresponds to a very high risk aversion, so high that no scenario with a nuclear component is a minimum cost scenario. The capital expenditure associated with a moratorium is large because nuclear plants under construction or recently commissioned must be paid for but never generate electricity, and substitute facilities must be built. The cost is so large that it compares with the total capital cost of the system. In our model's world, it is generally cheaper to endure an occasional accident, if occasional is infrequent enough. However, the real world may not choose the option with the lowest cost.

Terrorism and Sabotage—Safeguards

Here, one might expect an analysis of the high but improbable costs of the consequences of a successful terrorist act. Our model does not deal directly with the cost of such a possibility. In its place, we estimate the cost of safeguards adequate to reduce the likelihood of a terrorist act to an acceptably low level; this appears to be feasible.[19,20] If one wishes to include the costs of the terrorist act explicitly, one can add one's estimate of the probability-weighted cost to the cost of accidents.

One of society's decisions must therefore be what level of safeguards it is prepared to buy. Costs of safeguards in year t, $TS(t)$, are assumed to be proportional to the number of plants in service, $N_i(t)$. An additional cost is associated with safeguards during construction, which is taken to be proportional to the amount of new capital equipment, $\Delta N_i(t)$.

$$TS(t) = \sum_i TS_i N_i(t) + TS_i' \Delta N_i(t) \qquad (3-13)$$

CONCLUDING REMARKS

All costs with the same functional form affect the total cost in the same way. Fuel costs and property damage, for example, are proportional to the number of plants in service, and thus a higher assessment of the costs of property damage is equivalent to an increment to fuel costs. In this sense, ours is an abstract optimization model in which only four categories of costs are introduced; one could impute a cost for each category without specifying how it is apportioned within that category. Such a synthetic approach would not illuminate the effects of individual social costs as our empirical tabulation of costs does. Nonetheless the use of a small number of cost categories permits the planner to examine tradeoffs among the costs in a common category without having to reexamine the optimization of the model. Some costs cannot be assigned with certainty to a particular category so that it behooves us to watch the behavior of the larger costs to be sure they are categorized correctly. Direct fuel prices, spent fuel storage, and research and development are the most important of these costs.

Almost all our possible worlds have costs that fall into three categories, as indicated in the last two columns of Table 3−1. Category 1 costs depend on the incremental capacity of plants of type i, $\Delta N_i(t)$, in year t. Costs in Category 2 depend on $N_i(t)$, the total capacity of type i in operation in year t. Category 2 includes both those costs that depend only on total capacity in use, called Category 2a costs, and Category 2b costs, that depend both on $N_i(t)$ and on $\eta_i(t)$, the total energy generated between the beginning of the period and the year t by plants of type i. Category 3 includes the costs that depend only on $\eta_i(t)$. A fourth category, costs increasing linearly with time, enters in only one variable and is used in only a few scenarios.

Two kinds of costs have not been included in our model—the costs of regulation and other managerial costs that are not already included in the costs of research and development, and costs associated with strategic values of particular fuels. The latter cost is less relevant to electric power generation than to sectors in which oil and gas must be used. One might want to estimate strategic values and costs from the amounts that are spent on stockpiling, for example; this might have a significant effect on the cost of uranium.

The costs of regulation could be incorporated as follows. With the exception of the Nuclear Regulatory Commission (NRC), the activities of federal and state regulatory agencies have costs that are approximately independent of the mix of fossil and nuclear power plants. One could estimate what portion of the budget of the NRC goes toward the regulation of nuclear power plants and add that cost to the total. Presumably, the cost of regulation has a fixed annual base cost and additional costs associated with both the number of plants under construction and the number of plants in operation. Hence the annual regulatory cost should have a constant term, a term proportional to ΔN_n and a term proportional to N_n. One could assume that the other regulatory costs are the same for all mixes and therefore do not affect the choice.

※ *Chapter 4*

Analysis

DESCRIPTION OF SCENARIOS CONSIDERED

How sensitive is the optimum mix of power plants to the quantities and prices imputed to the variables just discussed? We examine conditions that represent likely bounds of possibility in the hope that the range of realistic cases is covered.

The analysis treats the United States for the thirty-year period 1975−2004. All costs are cited in constant 1974 dollars, discounted at an inflation-free rate, and summed over the period. There are two reasons for limiting the analysis to thirty years. First, even if the discount rate is modest, costs incurred far in the future have little effect on decisions made today. At a discount rate of 0.05, the present value of a cost incurred in thirty years is 23 percent of an equal cost incurred today. Second, if we knew what types of plants could be built in the next century, their costs would be difficult to predict.

Social costs are sensitive to the rate at which the future is discounted. The sensitivity of the optimal fossil to nuclear ratio to this factor is tested by calculating the costs with discount rates from 0 to 20 percent. A base case rate is set at 10 percent. In addition, calculations are done with and without discounting costs to human health and safety over the thirty-year period of the model. This decision is a much debated political and social one[21] that affects the total cost of the power-generating system.

The model is based on the assumption that enough electric power will be generated to meet an exogenous demand. Total costs have been evaluated with several forms for the temporal behavior of demand. All cases start from 1975 levels. Two choices assume exponential growth, one at 4 percent and the other at 7 percent per year. A 7 percent growth rate is the historical rate of annual increase in the U.S. electrical demand from the 1940s until about 1973 and is the highest growth rate we consider. A rate of 4 percent is based on slower recent growth trends and present price levels. This rate is used for base case calculations. The third case begins with exponential growth at 4 percent for fifteen years and then reverts to logarithmic growth. This case simulates a reduction in demand growth such as might occur in response to high electricity prices. The fourth choice is a linear increase beginning at the present growth rate. These four models are shown in Figure 4-1.

All power plants are assumed to have thirty-five-year lifetimes over which their output is constant. Lifetimes between thirty and forty years are projected in the Federal Power Commission National Power Survey[22] and Project Independence Report.[23] An effective lifetime of less (more) than thirty-five years would be equivalent in our results to an increase (decrease) in consumer demand to be satisfied by new plants. It is thus implicit in the calculation. The model begins with the plants in or about to go into service in 1974.[24]

The plant lifetime is not a simple number because a unit typically spends a portion of its life as a baseload unit and then serves as an intermediate or peak load unit as more efficient units come into service. But the total social cost depends only on the sum of the outputs of many plants. Because the number of plants is large and they are retired regularly, one can make the approximation that the output of each plant is constant over its lifetime. This assumption simplifies the calculations considerably.

For units introduced during the first fifteen years, only two fuels, coal and uranium, are considered. We assume that oil is too expensive to be a practical alternative to coal or uranium in the new units. The potential for conventional hydroelectric power is insufficient to supply large portions of growing demand. Prior to the 1977 Clean Air Act Amendments, coal plant operators had the options of burning high-sulfur coal with scrubbers of low-sulfur coal without; the preference was determined regionally on the basis of the relative prices of coal and of scrubbing. The 1977 amendments have the effect of requiring scrubbers for all new plants. Uranium is assumed to be "burned" in light water reactors (LWRs) because that is the principal type of operating reactor in the United States today. Pressurized and boiling water reactors are not distinguished because they have very similar costs.[6, 25]

Analysis 29

Figure 4-1. Consumer Demand vs. Time. These are, from the highest downward: 1, 7% exponential growth; II, 4% exponential growth; III, 4% annual growth through year 15, followed by logarithmic growth (with curve and derivative continuous), and IV, linear growth with initial slope equal to that of the 4% growth curve.

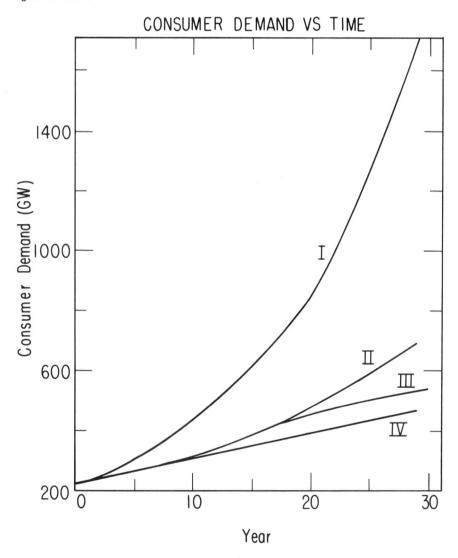

Two scenarios of new plant construction lead to the exclusive use of presently commercial technologies. In the first, coal plants and LWRs are employed for the period in question; research is done only to improve these technologies, and none is carried out on advanced technologies. Any constant mix from 0 percent nuclear (all coal) to 100 percent nuclear can be built.

In the second kind of scenario, research is carried out until 1990, at which time the advanced technology is declared a failure, meaning it is at best not yet commercial. Any mix can be built up to that time, but only coal plants can be constructed after 1990. The assumption is made that uranium has become prohibitively expensive and pollution problems from coal burning have been solved. The difference between the cost of this scenario and the first is the cost of research failure; the difference between this and the one described below is the result of successful research.

The third research scenario is an optimistic upper bound that assumes successful development of an alternative technology ready for commercial operation by 1990. In the sixteenth year, 1991, the decision is made to construct half the new plants with the new technology. From then on, the fraction of new technology plants rises linearly until the last year, when it is unity. The other plants built during the thirty years can be any mix of coal and nuclear power. For any calculation, the mix of fossil and LWR plants is assumed to stay the same after introduction of new technology as it was previously. The costs of new technologies were estimated for breeder reactors and central station solar facilities. Other possibilities are not excluded if their costs presumably correspond roughly to one or the other of these alternatives. The program used to evaluate the total costs is described in Appendix 2.

COSTS CALCULATED FROM THE MODEL

Two kinds of results emerge from our calculations. The first are the results that describe the costs of the scenarios we have considered and the determination of the least cost mixes for these scenarios. The second are the results that provide the reader with a simple, quick tool for introducing different or additional costs and for determining the optimal mix with those costs instead of the ones in our scenarios. This section presents the results obtained from our calculations and then describes how to use Tables 4–1 to 4–17 to extend the calculations to other situations and cost schedules.

With so many variables, it is easy to generate far more information than an ordinary person can assimilate. One is inevitably faced with reducing the data to a manageable form. We have done this in several ways. The simplest in its conception is a graph of the yearly cost as a function of time for a completely specified option. Figure 4-2 shows such curves as the total costs build up over the thirty-year period for four cases, all based on a consumer demand that grows at 4 percent per year, a discount rate of zero, and "base case" costs, as described in Appendix 1. The four cases, in order of decreasing cost, are: (1) all coal plants that are replaced by a successful new technology, and a high rate of escalation for coal costs; (2) all coal plants, no new technology, and no fuel cost escalation; (3) all coal plants initially, successful new technology, but no fuel cost escalation, and (4) all nuclear plants initially, successful new technology, and no fuel cost escalation.

Graphs such as Figure 4-2 give a simple picture of particular scenarios. The area under each curve through the thirtieth year tells the total social cost of that option. One could look at many such curves and sort out the lowest cost option. However, the data can be put in other forms that address more directly the questions of the composition of the minimum cost mix and the sensitivity of the composition of the optimum mix to individual costs. One of these forms is graphical, as a sort of map; the other is tabular. The graphical representation is illustrated by Figures 4-3 to 4-12.

Map Representations: Capital Costs

Figures 4-3 to 4-11 can be thought of as views downward onto a surface in which the two axes of the figures represent the costs of two alternative (substitutable) factors, such as the cost of capital equipment for coal and nuclear plants, K_c and K_n. We adopt the convention of always putting the nuclear cost on the horizontal axis and the alternate coal-fueled cost on the vertical axis. Projecting up out of the paper is the total social cost. Each point on the plane corresponds to the total cost of the electric power generating system when the costs of all the factors are chosen. All but two of the costs are constant over an entire graph; those two constitute the variables for that graph. In such a figure, it is natural to draw contours of constant height, curves, or lines along which the total cost of the optimal system is constant. Such contour maps are well known in the representation of production systems, that is, isoquants and isocosts. In our graphs, the contours are isocost or constant cost curves and are shown dashed in the figures.

The isocost curves shown in Figures 4-3 to 4-11 are identified by their total social costs, not simply by the costs of the two substitutable factors shown on the axes of the graphs. Each map is drawn for

Figure 4−2. Growth of Costs Through Time for Four Scenarios. All are derived from base case (Chicago Commonwealth Edison) factor costs, with 4% growth of demand and zero discount rate. In order of decreasing cost, they are: I, all coal plants, being replaced by successful new technology, and a high ($\beta = 0.0003$) rate of escalation of fuel costs; II, all coal plants, no new technology, and no change in fuel price; III, all coal plants, successful new technology, and no escalation of fuel prices; IV, all nuclear plants, successful new technology, and no fuel cost escalation.

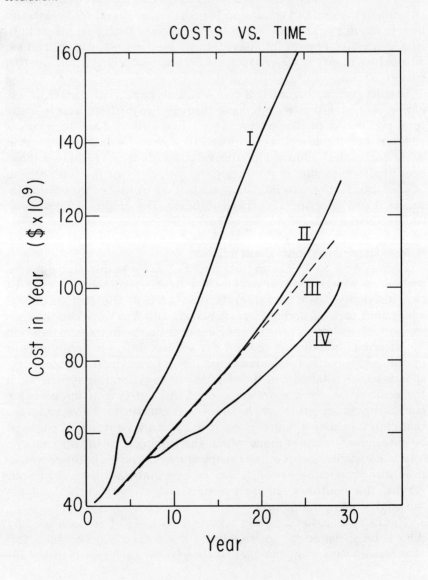

Analysis 33

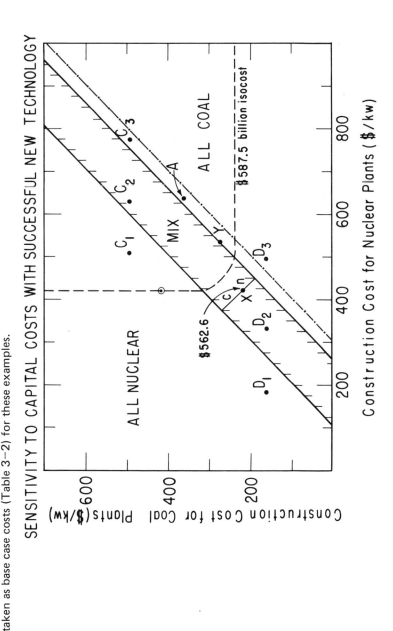

Figure 4–3. Map of the Total Social Cost Surface with Construction Costs K_c and K_n as Variables. Demand grows at 4%, the discount rate is 10%, the nuclear fuel cost escalator is 2×10^{-4}, and the coal cost escalator is 2×10^{-5}. Successful new technology is assumed. The open circle on the $587.5 billion isocost curve corresponds to base case (Chicago Commonwealth Edison) costs as given in Table 3–2. The point A corresponds to capital costs of the Alabama Power Company; points C_1, C_2 and C_3 correspond to figures for Cleveland Electric, and points D_1, D_2 and D_3, to the Duke Power Company, as described on p. 35. Other costs are taken as base case costs (Table 3–2) for these examples.

a particular assumed demand and for specific discount rates for human life and for all other factors.

On each map, the "base case" is shown as a circled point. These costs are based on the Chicago region; direct costs are those of the Commonwealth Edison Company for 1974. It is possible to evaluate other specific situations and regions without reference to a standard base case, and we shall describe results for several regions in discussing the maps. However the introduction of a base case simplifies the use of the model as a quantitative tool, as we shall see at the end of this section.

The space of each map can be partitioned to correspond to three kinds of solutions to the optimal mix problem. Where the cost of a nuclear system is high and the corresponding cost of a coal-fueled system is low, the optimum is achieved by using only coal-fired plants. Likewise, the region where the coal-fueled systems are expensive and nuclear plants are inexpensive is the region in which the optimum mix is all nuclear. For the most simplistic scenarios, the only choices are all fossil or all nuclear, the so-called corner solutions. The dash-dot boundary of Figure 4-3 is an example discussed below. For scenarios with nonlinearities, the possibilities are richer as the other partitions of Figure 4-3 will illustrate. If, for example, fuel prices escalate, the lowest-cost combination includes some fossil and some nuclear plants. Along any isocost curve, the composition of the optimum mix varies smoothly through the mixed-source region from all of one kind of source to all of the other.

The mix region is separated from the all-fossil and all-nuclear regions by straight line boundaries. Whenever the map represents variables for which the cost depends on ΔN or N, the boundary lines of the mix region are parallel, as in Figure 4-3. (This is not true for other factors such as fuels with escalating costs, as we shall see.) To evaluate the composition of the optimum ratio of fossil to nuclear plants for a point lying in the mix region, one draws a line through the point perpendicular to the mix region boundaries and measures the lengths of the segments from the point to the boundaries. The optimum mix has the two kinds of plants in the ratio of the two segments. For example, in Figure 4-3, the optimum mix at point X consists of a fraction $c/(c+n)$ of coal-fired plants and $n/(c+n)$ of nuclear-powered plants, where c and n are the lengths shown in the figure.

With more alternatives, particularly with the possibility of new technology, the number of distinguishable areas on a map increases. In Figure 4-9, there are six different areas, corresponding to all fossil-fueled plants, fossil with a transition to new technology, a mix of fossil and nuclear plants, a mix of fossil and nuclear phasing into new technology, all nuclear plants, and finally, all nuclear plants phasing into new technology.

To illustrate how to read and interpret these maps, consider Figure 4-3. The circled point represents the base case corresponding to capital costs of 421×10^6 for K_n, the nuclear cost, and for K_c, for coal-fired plants with scrubbers. The demand is assumed to grow 4 percent annually, the discount rate is 10 percent, and other costs are as given in Table 3-2 as "best estimates." The nuclear fuel escalator is 2×10^{-4}, and the coal cost escalator is 2×10^{-5}. For this map, we assume that new technology—taken here to be breeder, rather than solar, but the difference is small—is developed successfully and is introduced in the sixteenth year as described in the previous section. The base case point lies in the "all-nuclear" region, near the border of the "mix" region, on the $587 billion isocost contour.

If the price of nuclear plants were lower than $421 million per 1,000 Megawatt (MW) plant or if the price of coal-fired plants increased above $421 million, the lowest cost solution would remain 100 percent nuclear. However, if the capital cost of nuclear plants remained constant but the capital cost of a coal-fired plant dropped to $220 million, the minimum cost solution would fall at the point X on the $562 billion isoquant and in the "mix" region. In fact, the optimum mix for this situation would be about one-third nuclear and two-thirds fossil. We obtain that ratio from the ratios of lengths $c/(c + n)$ and $n/(c + n)$ as indicated previously. If K_c were to drop below $165/kw while K_n remained at $421/kw or K_c remained fixed at $421/kw and K_n went above about $680/kw, then the lowest total cost system would be entirely fueled by coal. This would also be the case if the cost of coal-fired plants were $275/kw or less and the costs of nuclear plants were $535/kw or more, as at point Y of Figure 4-3.

Such cost estimates are not unrealistic. Point A in the all-fossil area is based on the data quoted in the 1975 Federal Power Commission Report[22] for the capital costs estimated by the Alabama Power Company for construction in 1977-1980. Point D_1 is based on the figures quoted for the Duke Power Company with K_n for 1973-1974 and K_c for 1973-1976 (the only period for which Duke's costs of coal plants are quoted); D_2 corresponds to the same K_c but K_n for 1977-1980; and D_3 to the same K_c but to K_n estimated for 1981-1982. We cannot pretend that K_c for 1981 will be the same $162/kw that it was in 1973, and thus points D_2 and D_3 must be taken only as representatives of hypothetical cases rather than real. Taken literally, the progression from D_1 to D_2 to D_3 would indicate that the capital costs favored an all-nuclear commitment in 1973 but that the cost estimates of Table 3−2 would swing the mix entirely to coal-fired plants in the future for that geographic region. The same happens with the capital cost estimates of Cleveland Electric, corresponding to points C_1, C_2, and C_3, with values of K_n for 1977−1978, 1979−1983, and 1983−1984, respectively, and K_c an average for 1975−1980.

Thus far, we have discussed Figure 4−3 with regard to the case in which fuel costs escalate so that there is a region in which the optimum is neither all nuclear nor all coal-fueled plants. One can consider at the same time the simpler case in which fuel costs remain constant and no other costs depend on the accumulated fuel consumed. For such a situation, the only solutions are all nuclear or all coal-fueled plants. The dividing line is the dash-dot line of Figure 4−3.

The model of Figure 4−3, like most of the succeeding maps, is drawn under the assumption that new technology with breederlike costs is developed successfully and introduced. The effect of the costs of new technology is to raise all the isocost lines without changing their shape. For this case, in which the discount rate is 10 percent and demand grows at 4 percent per year, the increment to each isocost line is approximately $6 billion for every $100/kw increase in the cost of new technology. Hence the difference between $500 and $1000/kw, two of the estimates of capital costs for the breeder, means a difference of $30 billion in the labels on isocost lines, about 5 percent of the total cost for the $589 billion isocost line of Figure 4−3. Figure 4−3 isocosts are based on $600/kw for new technology. If one goes to a solar model for estimating costs of new technology, the isocosts should be labeled $24 billion higher than shown in Figure 4−3 to correspond to an estimate of $1000/kw for K_p of solar electric systems.

Figure 4−4, with K_n and K_c as the variables once more, is based on the same cost premises as Figure 4−3 save one. The map of Figure 4−4 is constructed on the basis of a 7 percent annual growth of consumer demand rather than 4 percent. Because much more electricity must be generated, the effects of fuel cost escalation are more telling than for the situation in Figure 4−3. The "mix" region is much broader, and the Chicago-Commonwealth Edison base case point now lies in this mix region with a best mix of approximately 25 percent coal and 75 percent nuclear. Alabama Power's best choice, if all other variables were the same as they are for the base case, would still be all coal plants but just marginally (point A). Both Duke and

Analysis 37

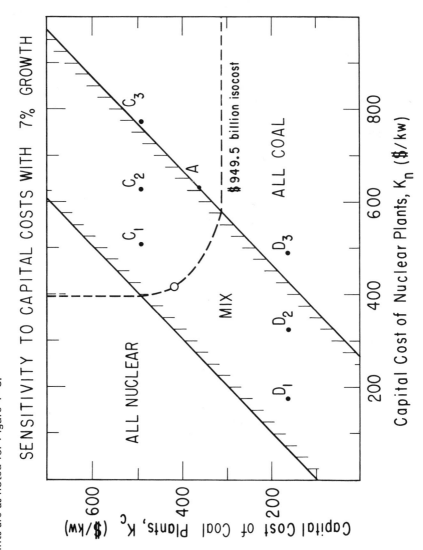

Figure 4-4. Map of the Total Social Cost Surface with Construction Costs K_c and K_n as Variables. All other variables are taken as for Figure 4-3 except that demand is assumed to grow at 7% per annum exponential rate. The open circle is the "base case" and other points are as noted for Figure 4-3.

Cleveland Electric would optimally choose a mix of fossil and nuclear plants (points D_1, D_2, C_1, C_2) so long as their capital costs for coal plants remain constant and their nuclear capital costs stay in the range of their lower estimates. The absolute capital costs estimated by Duke Power and by Cleveland Electric are very different, but with the other costs fixed as they are for Figures 4-3 and 4-4, these two power companies would reach similar decisions.

The effect of changing the discount rate is illustrated by Figures 4-5 and 4-6. Again, the capital costs K_n and K_c define the axes. Figure 4-5 is based on a zero rate of discount so its total costs are very high, $2173 billion for the base case. Figure 4-6 has been computed for a 20 percent discount rate, and the base case total cost is relatively low, $294 billion. The mix region is narrow when the discount rate is high because fast discounting wipes out the effect of the fuel price escalator in the distant future when it would be most important. Alternatively, one can take the view that fast discounting minimizes the effect of scarcities or surfeits many years hence. For a low or zero discount rate, nuclear fuel or a mix is favored. For a high discount rate, where the mix range is narrow, the decision is a sensitive function of capital costs, as the locations of the points A, C_1, C_2, C_3, D_1, D_2, and D_3 indicate.

Low growth rates, either linear or exponential to logarithmic, have little effect on the optimum mix relative to that of the base case (Figure 4-3), except to shift it toward the fuel with the more rapidly escalating price. If we continue to take the nuclear fuel escalator as ten times that for coal, as for the maps of Figures 4-3 to 4-6, then the map for linear growth (starting with 4 percent per annum) takes the form of Figure 4-7. The Alabama Power choice is again barely all fossil, Duke Power chooses as in Figure 4-3, Cleveland Electric swings a bit more toward nuclear, and Commonwealth Edison becomes more firmly all nuclear than before. Were we to guess that coal prices will increase faster than uranium prices, the shift would be in the other direction.

Figure 4-8 is based on the same assumptions as Figure 4-3 except that no research is done, no new technology is introduced, and the same mix of plant types is built throughout the thirty years. This figure is very much like Figure 4-3, partly because the discount rate of 10 percent makes the last few years of the thirty-year period contribute relatively little to total cost, partly because the research costs are small, and partly because these illustrative maps were drawn for the case of a small increase in the price of coal. We shall return to the question of new technology shortly.

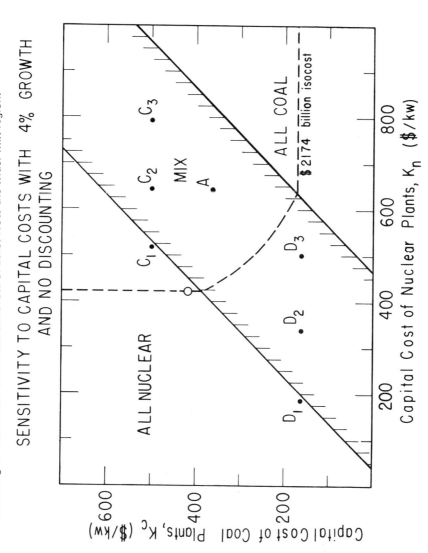

Figure 4–5. Map of the Total Social Cost Surface with Construction Costs K_c and K_n as Variables. This map differs from Figure 4–3 in that this figure was drawn with a discount rate of zero. Note the wider mix region.

40 TOSCA

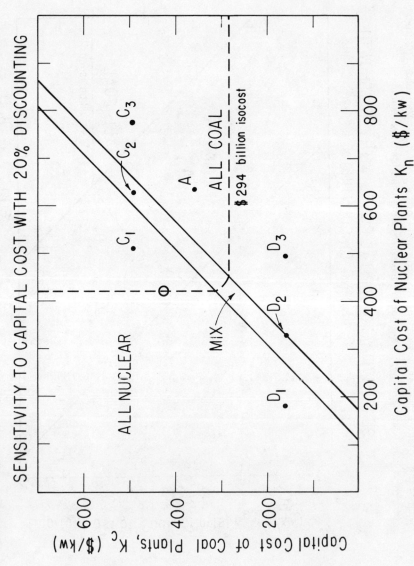

Figure 4-6. Map of the Total Social Cost Surface with Construction Costs K_c and K_n as Variables and a Discount Rate of 20%. All other variables were selected as for Figure 4-3. High discounting has the effect of narrowing the mix region.

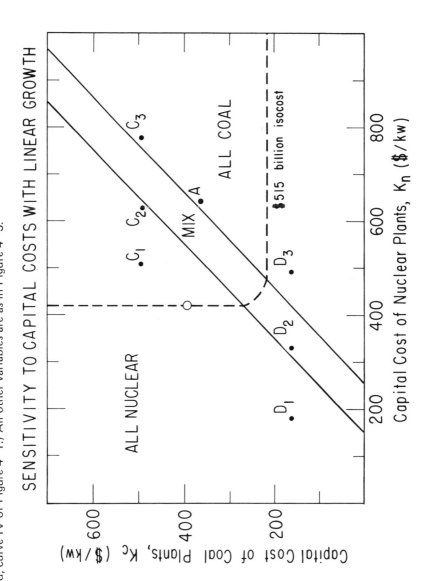

Figure 4-7. Map of the Total Social Cost Surface with Construction Costs K_c and K_n as Variables. (Based on a linear growth of demand, curve IV of Figure 4-1.) All other variables are as in Figure 4-3.

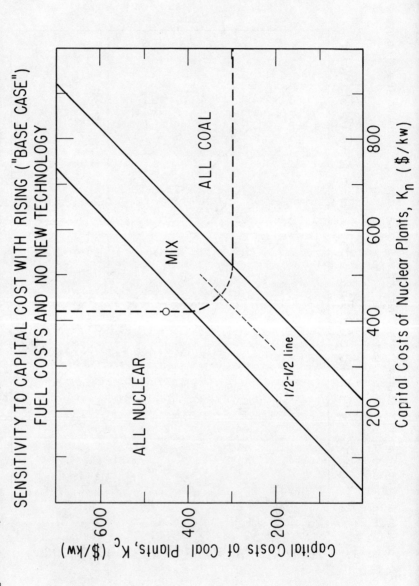

Figure 4-8. Map of the Total Social Cost Surface with Construction Costs K_c and K_n as Variables. This map is drawn assuming that no new technology is developed or introduced, that fuel costs escalate as given for the base case ($\beta_n = 2 \times 10^{-4}$, $\beta_c = 2 \times 10^{-5}$), that demand grows at 4%, the discount rate is 10%, and all other variables are as shown for the base case in Table 3-2.

Map Representations: Human Costs

Next we examine the effects of human costs—health effects and accidents—on the optimum mix. Figure 4-9 has as its axes the human costs of nuclear power and the human costs of coal-fueled power. Public concern has been dominated by the potential large accident, whose annual costs have the form of:

(cost of a human life)

× (probable number of lives lost in an accident)

× (probability of an accident in year t)

× (discount factors for year t)

× (risk aversion factor)

Figure 4-9 actually contains the representation of two alternatives— the solid parallel diagonals bound the mix region when human costs are not discounted, and the dashed parallels define the narrower mix region when human costs are discounted at the same 10 percent rate as all other factors in the scenario. The total cost at any point in this map depends on how we discount. The base case point, far into the all-nuclear region, corresponds to a total social cost of $587.5 billion if human costs are discounted at 10 percent and to $672.1 billion if they are discounted at a zero rate.

The human costs from power generation are likely to be highly controversial. Two of the factors that would affect our estimate of the costs on the nuclear side are health effects to uranium miners greater than our estimate of $1 million or one death per year per plant and much greater expected mortality for accidents than the number of lost lives estimated in the Rasmussen Report.[26]

Let us consider accidents. If we take the Rasmussen Report's estimated number of lost lives, 50,000 (including long-term effects such as latent tumors and genetic defects), attribute $50 billion to the direct human costs, and add $14 billion for property damage, the $64 billion represents the cost estimate of the worst plausible accident. The Rasmussen Report assigns a probability of one part in 10^9 per reactor year of such an accident, which implies a probable human (plus property damage) cost of this accident of $64 per reactor year. It is clearly negligible compared with the other human costs. The integrated probable cost per year of all accidents is $50,000. To begin to affect the position of the base case point, and thus the decision, the probable accident cost per reactor year, multiplied by risk aversion, must be at least 25 times larger than the figure of $50,000. If the accident cost were $1.28 million, the point would shift one and

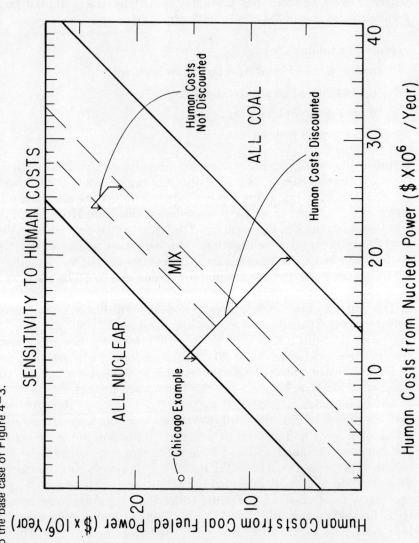

Figure 4–9. Map of Total Social Costs with Human Costs H_c and H_n as Variables. The solid diagonals define the mix region when human costs are discounted; the dashed diagonals define the mix region if human costs are not discounted. Other factors correspond to the base case of Figure 4–3.

one-fourth units to the right in Figure 4−9 and still fall well in the all-nuclear region whether or not human costs are discounted. Note that 25 is the magnitude of the factor by which the Rasmussen figures must be challenged in order to move the base case perceptibly, a factor of about 250 is required to shift into the mix region if human costs are discounted, and a factor of over 400 is required to move the system into the all-fossil region if human costs are not discounted.

One can challenge a probability of one in 10^9 for the worst accident; whether the total accident rate can be multiplied by 100 will depend not on small corrections but serious challenges to the methods used in the risk assessments. Part of the factor might be attributed to risk aversion associated with social unacceptability of a large accident. However some caution must be exercised by one who would introduce a risk aversion factor for one human cost and not another; it necessarily implies a choice reflecting a social preference. If one attributes a risk aversion to nuclear accidents, then one must be prepared to find strong justification for attributing any other factor to accidents and lives lost in connection with the coal-fueled cycle.

The cost of insurance for nuclear facilities, discussed in Appendix 1, is negligible under the Price-Anderson Act, but it should be sufficiently higher than the direct costs of life and property to provide income to the insurers. This means that insurance costs can be larger but not very much larger than the direct accident costs.

The human cost to uranium miners may be underestimated in the base case. This factor is likely to become better known and more soundly estimated as people work for extended periods in uranium mines and their health records become available. However recall that even a hundredfold increase in the human costs from a large nuclear accident is not itself enough to justify a shift away from the base case choice of 100 percent nuclear power. Other cases closer to the borderline can be affected by significant but reasonable reevaluations of the human costs.

There is a substantive question as to whether costs to human health should be discounted:

> There have been long-standing debates as to the appropriateness of applying a discount rate to effects on future generations, since any positive rate of discount will directly discriminate in favor of choices that involve bad impacts on later generations but not on earlier ones. If the discount rate were 5 percent, 100 cases of toxic poisoning 75 years from now would be equivalent to about 3 cases today; or 1 case today would be valued at the same as 1730 cases occurring in 200 years. . . . Clearly, intergenerational effects of these magnitudes are ethically unacceptable.[21]

46 TOSCA

If human costs are not discounted, the mix region narrows, and therefore higher estimates of human costs per reactor year are required to move the optimum choice away from the all-nuclear option. However if the mix region is narrow, only a small increment carries the minimum cost system over to the all-coal region.

Map Representations: New Technology

In order to exhibit the effects of the capital costs of new technology on the policy of lowest total social cost, we turn to Figure 4–10, a slightly more complex map than the others. Recall that in the first section of this chapter we chose a specific scenario for the introduction of this new technology—50 percent of the new power plants in the sixteenth year use new technology, and the proportion of new technology plants rises linearly until all the new plants built in the thirtieth year use the new technology. The two substitutable options here are new technology and coal-fired plants. Corresponding to these options, the axes represent the capital costs K_ν for plants with new technology and K_c for coal-fueled plants. There are six possible options, corresponding to the six regions a to f on the map of Figure 4–10. The Chicago Commonwealth Edison case, with K_ν estimated as \$600/kw, has a lowest cost option corresponding to all nuclear plants to begin and new technology to replace them. The total cost is, of course, the base case cost of \$587.5 billion. With the same estimate of K_ν but the coal plant capital costs K_c quoted for Alabama Power (point A) or for Cleveland Electric (point C), the decision is the same, that is, start with all nuclear plants and, beginning in the sixteenth year, introduce plants with new technology. Duke Power (point D), by contrast, has such a low (1973) value for K_c that, if all the costs for that utility were close to those of our base case, it would be best, in the sense of lowest total social cost, to build only coal-fired plants. If the cost of coal plants were increased to \$200/kw, then Duke Power should construct a mix of coal and nuclear but introduce no new technology. For the utility to adopt new technology, its K_c would have to double or K_ν would have to decrease. The cost of capital for new technology would have to be less than \$230/kw to force all utilities to choose that option.

Map Representations: Fuel Costs

Figure 4–11, the final map, shows the effects of the costs of the two fuels. With prices escalating at the base case rates, one obtains the mix region bounded by the solid converging lines with an all-nuclear region above and an all-coal region below. With no price escalation, the all-nuclear and all-coal regions are bounded by the

Analysis 47

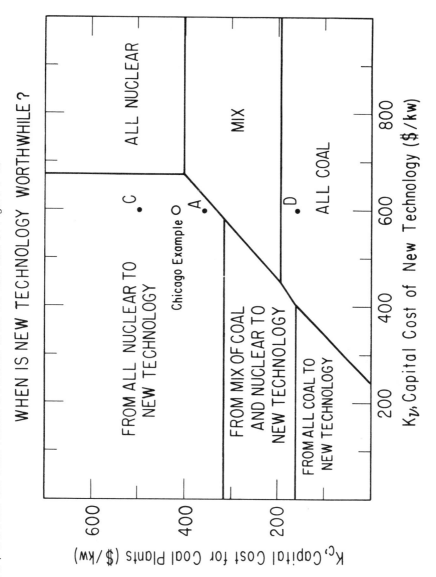

Figure 4-10. Map of Total Social Costs with Capital Costs of Coal Plants K_c and of New Technology K_ν as Variables. Nuclear plant costs and other factors are chosen as for the base case of Figure 4-3.

Figure 4-11. Map of Total Social Cost Surface with Initial Fuel Costs as the Variables. If fuel costs escalate according to the base case model, the mix region is bounded by the converging solid diagonals. If fuel costs do not escalate, there is no mix region, and the all-coal and all-nuclear regions meet at the dash-dot line. All other factors are those of the base case of Table 3-2 and Figure 4-3.

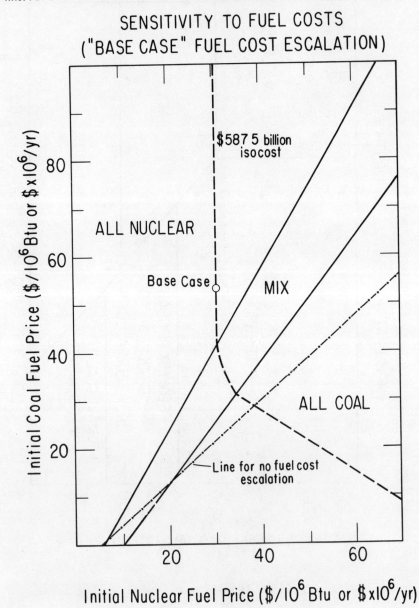

dash-dot line. The base case point lies in the all-nuclear region; the $587.5 billion isocost is shown. If the cost of uranium were increased from the base case assumption of $31 million/year to $37 million/year and coal costs remained constant, an all-nuclear program would no longer be optimal. If the cost of uranium rose above $52 million per year for a 1,000 MW reactor, the least cost choice is all coal. This would correspond to a total cost of $631 billion at the first all-coal point. This is telling us that the choice of optimal mix is very sensitive to the relative prices of coal and uranium. Both of these are difficult to predict; uranium prices are highly uncertain. There may be enough public benefit to justify some public expenditure on uranium exploration to help reduce this uncertainty.

Tabular Representations

The graphs and maps give us a qualitative picture of the choice implied by any chosen set of parameters. To make the interpretation more quantitative and to exhibit the effects of several factors together, we put the results in tabular form. Tables 4–1 to 4–4 give total costs for mixes ranging from no nuclear power through all-nuclear power for four different forms of consumer demand. They include possibilities that neither or either one of the fuel costs may escalate. These tables all suppose a transition to new technology and exhibit costs for discount rates of 0, 5 percent, 10 percent, and 20 percent. (Tables 4–5 to 4–9 are incremental cost coefficients to be used with Tables 4–1 to 4–4 as discussed in the next section.) Tables 4–10 to 4–13 are counterparts of the total cost Tables 4–1 to 4–4 for the situation in which there is no research for new technology and therefore no new technology. To model failed research for breeder technology, we add the cost of research—$9.1 billion if it is stopped after year 15 and the discount rate is 10 percent—to the costs in Tables 4–10 to 4–13. Tables 4–14 to 4–18 are additional tables of incremental cost coefficients appropriate to Tables 4–10 to 4–13.

To illustrate the use of these tables, we turn first to Table 4–1, which assumes a zero rate of discount. If consumer demand grows at 4 percent per year and neither fuel cost changes, we see that an all-coal program costs $2252 billion, a 50 percent-coal, 50 percent-nuclear program costs $2120 billion, and an all-nuclear program has the lowest total social cost of $1988 billion for this set of conditions. If fuel costs do not escalate or if coal costs escalate and nuclear costs do not, the lowest cost option is nuclear for all four demand scenarios. However, if nuclear fuel costs escalate with a factor as small as 0.0003/year and coal costs are constant, the system with lowest total

Table 4–1. Total Social Costs: Base Case, with discount rate of zero, successful new technology, and human costs not discounted. Total costs are in billions of dollars ($ $\times$ 10^{-9}). Lowest cost options are underlined.

Fuel Cost Escalation	Mix (fraction nuc) \ Consumer Demand	4%	7%	Exponential to Log	Linear
0	0	2252	4422	1985	1761
	1/4	2186	4261	1923	1715
	1/2	2120	4100	1861	1670
	3/4	2054	3940	1798	1624
	1	_1988_	_3779_	_1736_	_1578_
Nuclear Fuel Escalation = 0.0003	0	2269	4439	2003	1778
	1/4	2239	_4399_	1973	1755
	1/2	_2228_	4476	_1962_	1742
	3/4	2238	4670	1968	_1738_
	1	2267	4981	1992	1743
Fossil Fuel Escalation = 0.0003	0	2831	6292	2535	2203
	1/4	2649	5450	2370	2102
	1/2	2501	4804	2234	2016
	3/4	2385	4352	2128	1947
	1	_2303_	_4095_	_2051_	_1894_

costs is a mix whose composition may range from 25 percent to 75 percent coal-fired, depending on the demand.

Tables 4–2 to 4–5 illustrate the effect of discounting future social costs. Let us use Table 4–3, with its 10 percent discount rate, to illustrate this added complexity. If all costs are discounted, fuel costs do not escalate, and demand grows at 4 percent, the total social cost is $609 billion for an all-coal system and $563 billion, the lowest of that set, for an all-nuclear system. If we exempt the human costs from discounting, the corresponding figures are $784 billion for all-coal and $648 for all-nuclear systems. Whether or not we discount human life, with no fuel cost escalation or with coal costs escalating

Table 4-2. Total Social Costs: Base Case, with discount rate of 5% and successful new technology. Boxes divided diagonally (zero fuel cost escalation portion) give total costs for discounted human costs (upper figure) and for undiscounted human costs (lower figure). All other boxes give figures with human costs discounted. To obtain total costs when human costs are not discounted, follow the procedure given on pp. 57. Total costs are in $billions ($ × 10⁹); lowest cost options are underlined.

Fuel Cost Escalation	Mix (fraction nuc) \ Consumer Demand	4%	7%	Exponential to Log	Linear
0	0	1063 / 1189	1881 / 2122	983 / 1007	884 / 984
	1/4	1038 / 1146	1818 / 2018	958 / 1060	866 / 954
	1/2	1012 / 1104	1754 / 1914	933 / 1020	847 / 925
	3/4	986 / 1062	1691 / 1810	908 / 980	829 / 895
	1	_960_ / _1019_	_1628_ / _1707_	_884_ / _940_	_811_ / _866_
Nuclear Fuel Escalation = 0.0003	0	1070	1888	989	890
	1/4	1058	_1866_	976	880
	1/2	_1050_	1882	_969_	873
	3/4	_1049_	1936	_968_	_869_
	1	1055	2028	972	_869_

Analysis 51

Table 4-2. continued

Fuel Cost Escalation	Mix (fraction nuc)	Consumer Demand	4%	7%	Exponential to Log	Linear
Fossil Fuel Escalation = 0.0003	0		1301	2541	1212	1077
	1/4		1237	2255	1153	1040
	1/2		1185	2033	1104	1009
	3/4		1143	1874	1065	983
	1		*1112*	*1779*	*1035*	*963*

Table 4–3. Total Social Costs: Base case, with discount rate of 10% and successful new technology. Human costs are discounted for figures in upper half-boxes and in undivided boxes and are undiscounted for figures in lower half-boxes, as described for Table 4–2. Total costs in $billion ($ × 10^9); lowest cost options are underlined.

Fuel Cost Escalation	Mix (fraction nuc) \ Consumer Demand	4%	7%	Exponential to Log	Linear
0	0	609 / 784	974 / 1304	583 / 748	533 / 673
	1/4	598 / 750	945 / 1220	571 / 715	525 / 650
	1/2	586 / 716	916 / 1136	560 / 683	517 / 626
	3/4	575 / 682	887 / 1052	549 / 651	508 / 602
	1	<u>563</u> / <u>648</u>	<u>859</u> / 969	<u>538</u> / <u>619</u>	<u>500</u> / <u>579</u>
Nuclear Fuel Escalation = 0.0003	0	612	977	586	536
	1/4	606	<u>964</u>	579	531
	1/2	<u>601</u>	<u>965</u>	575	527
	3/4	<u>599</u>	979	<u>572</u>	<u>524</u>
	1	<u>600</u>	1008	<u>572</u>	<u>523</u>

Table 4–3. continued

Fuel Cost Escalation	Mix (fraction nuc) Consumer Demand	4%	7%	Exponential to Log	Linear
Fossil Fuel Escalation = 0.0003	0	725	1243	696	633
	1/4	700	1133	672	618
	1/2	678	1047	652	605
	3/4	661	983	635	594
	1	_648_	_943_	_622_	_585_

Table 4-4. Total Social Costs: Base case, with discount rate of 20% and successful new technology. Human costs are discounted for figures in upper half-boxes and in undivided boxes and are undiscounted for figures in lower half-boxes, as described for Table 4-2. Total costs are in $billion ($ × 10^9). Lowest cost options are underlined; where one or more options are virtually equivalent to the lowest cost option, all are underlined.

Fuel Cost Escalation	Mix (fraction nuc) \ Consumer Demand	4%	7%	Exponential to Log	Linear
0	0	302 / 511	417 / 804	298 / 497	281 / 451
	1/4	299 / 481	408 / 732	295 / 469	279 / 431
	1/2	296 / 453	400 / 660	292 / 442	277 / 411
	3/4	292 / 423	391 / 588	289 / 414	274 / 390
	1	<u>289</u> / <u>394</u>	<u>383</u> / <u>517</u>	<u>286</u> / <u>386</u>	<u>272</u> / <u>370</u>
Nuclear Fuel Escalation = 0.0003	0	303	418	299	282
	1/4	301	413	297	281
	1/2	<u>299</u>	<u>410</u>	<u>295</u>	<u>279</u>
	3/4	<u>298</u>	<u>409</u>	<u>294</u>	<u>278</u>
	1	<u>297</u>	416	<u>293</u>	<u>277</u>

Table 4-4. continued

Fuel Cost Escalation	Mix (fraction nuc) / Consumer Demand	4%	7%	Exponential to Log	Linear
Fossil Fuel Escalation = 0.0003	0	344	487	340	321
	1/4	339	464	335	317
	1/2	334	445	330	314
	3/4	329	430	326	311
	1	_326_	_419_	_311_	_309_

and nuclear fuel costs constant, the lowest cost systems are all-nuclear. If we discount human costs, if nuclear fuel costs escalate (β of 0.0003), and if demand grows annually at 4 percent, the system with lowest costs is three-fourths nuclear but only barely so. Its cost is $599 billion, and the 50–50 mix and all-nuclear systems cost $601 and $600 billion, identical within the uncertainties of the calculation.

Suppose we want to evaluate total costs for a situation with a 10 percent discount rate, an escalating fuel cost, and undiscounted human costs. We continue to use Table 4–3, but we now must perform a subtraction and an addition to obtain those costs. The increments attributable to discounting human costs are the same for corresponding boxes under all three fuel cost escalations. We find the desired increment in the top part of the table and add it to the figure in the corresponding box in the lower part of the table. Let us illustrate with a 7 percent growth case and with escalating nuclear fuel cost. When human costs are discounted, the lowest cost mix is between 25 and 50 percent nuclear; these two figures tie, with $964 and $965 $\times$ 10^9. Now suppose we do not discount human costs. From the zero escalation matrix, we find that the increment of human costs is $1220–945 or $275 $\times$ 10^9 for 25 percent nuclear and $1136–916 or $220 $\times$ 10^9 for the 50–50 mix. With these increments added to $964 and $965, the 25 percent nuclear case costs $1239 while the 50–50 mix costs $1185. Thus the balance swings away from the 75 percent coal choice toward more nuclear power. Indeed, the lowest cost mix with undiscounted human costs is all nuclear, with a cost of $1118.

It must be emphasized that the numbers in these tables are derived for the Chicago Commonwealth Edison case, whose factor costs favor nuclear power. Thus we should not be surprised at the absence of any lowest cost solutions calling for 100 percent coal-fired plants in Tables 4–2 to 4–5 and 4–10 to 4–13. We have already seen that in other geographic regions, 100 percent coal scenarios can be the cheapest. The following section describes how tables like Tables 4–2 to 4–5 and 4–10 to 4–13 for the Chicago Commonwealth Edison case can be constructed readily for other regions.

What is the value of research on new technology? The analysis here is much less elaborate than that by Manne,[1] but it reveals some of the features of the rather complex answer to that question.

The cost of fifteen years of research in our estimates is about $9.1 billion for the breeder model of new technology when computed at a 10 percent discount rate. From Table 4–3 and Table 4–12, we obtain the costs of comparable programs based on this discount rate,

Table 4-5. Incremental Cost Factors for Category 1 Costs, which vary as the number of plants added per year, ΔN, when new technology is successful. Units are chosen so that multiplication of the incremental cost in $hundreds/kw or $hundred million/(plant) by the tabulated coefficient gives the total cost increment in $billions ($ $\times$ 10^9). (The tabular quantities are thus in units of 10^{-7} kw). For example an increase of $150/kw in construction cost, for a 4% growth of consumer demand and a 10% discount rate, will add 1.5 $\times$ 14 or $21 billion to the total 30-year cost.

Discount Rate (%)	Consumer Demand 4%	7%	Exponential to Log	Linear
0	43	106	60	29
5	24	58	22	16
10	14	35	14	10
20	11	17	6	45

Table 4-6. Incremental Cost Factors for Category 2a Costs, varying as $N(t)$ and with no escalation of fuel costs and successful new technology. The units are chosen so that the increment in unit cost in $millions per plant, multiplied by the appropriate coefficient from the table, gives the increment to the total 30-year cost in $billions ($ $\times$ 10^9). Thus if the taxes on power plants increased to $2 million per year, per plant, then for a 4% growth of electrical demand and a 10% discount rate, the total increment to the 30-year cost would be 2 $\times$ 1.3 or $2.6 billion.

Discount Rate (%)	Consumer Demand 4%	7%	Exponential to Log	Linear
0	7.3	17.7	8.6	5.1
5	3.4	8.3	2.7	2.4
10	1.3	3.3	1.3	0.9
20	0.4	1.0	0.4	0.3

Table 4-7. Incremental Cost Factors for Category 2b Costs, varying as $N(t)$ with fuel costs escalating ($\beta = 0.0003$) and successful new technology. Units are the same as those of Table 4-6.

Discount Rate (%)	Consumer Demand 4%	7%	Exponential to Log	Linear
0	12.3	47.3	11.3	7.5
5	4.5	16.7	4.2	2.8
10	1.9	6.8	1.8	1.2
20	0.5	1.6	0.5	0.3

Table 4-8. Incremental Cost Factors for Category 3 Costs, varying as $\eta(t)$, the cumulative number of plants or amount of electricity generated, with successful new technology. Units are chosen so that if the incremental unit costs are given in $million per plant-year, multiplication by the appropriate factor in the table will yield the increment to the total 30-year cost in $billions ($ × 10^9). For example, if the cost of perpetual care of spent nuclear fuel were to increase $1.5 million per plant-year of operation and demand were growing at 4% with a discount rate of 5%, the incremental 30-year cost would be 1.5 × 24 or $36 billion.

Discount Rate (%)	Consumer Demand 4%	7%	Exponential to Log	Linear
0	65	118	63	65
5	24	46	23	19
10	10	15	10	8.3
20	2.7	4.7	2.3	2.3

Table 4-9. Incremental Cost Factors for Changes in the Escalator β for Human Costs from Coal-Fired Plants. Factors are based on successful new technology and on the base case estimate of 16 lives per plant-year. Changes in the escalator given in units of 0.0001/plant-year, multiplied by the tabulated factor, give the total change in 30-year cost in $billions ($ $\times$ 10^9).

Discount Rate (%)	Consumer Demand		Exponential to Log	Linear
	4%	7%		
0	27	156	24	13
5	9	51	8	4
10	1	19	2.5	2
20	1	4	0.5	1

with and without new technology, respectively. For example, with no escalation of fuel costs and a 4 percent annual growth of demand, the all-nuclear, lowest total cost solution has a cost of $548 billion with no new technology and a cost of $563 billion with new technology, a difference of $15 billion. If demand grows faster (7 percent) and nuclear fuel cost escalates, the optimum mix invokes new technology. With 25 percent nuclear, the total cost of $964 $\times$ 10^9 is only $3 $\times$ 10^9 less than the $967 $\times$ 10^9 cost of a system with no new technology, but with a 50-50 mix, the $965 $\times$ 10^9 cost of a system with new technology is distinctly lower than the $985 $\times$ 10^9 for a system having no new technology.

The cost of a program whose research is unsuccessful is the cost of a system without new technology, plus the $9 billion for research. Suppose the $9 billion research cost is a mandatory add-on for any program that includes nuclear power. Then, we see from Tables 4-3 and 4-12 that with no escalation of fuel costs, the lowest cost solutions introduce no new technology (and therefore include no research for this technology). However, if nuclear fuel costs escalate or if research is successful, we indeed choose to adopt the new technology to reach the lowest cost program.

These inferences are weak in that they do not consider the costs beyond the thirty-year period. To make a firmer comparison, one would have to extend a study like ours to compare coal and nuclear plants with plants based on new technology for a period at least comparable with their useful lifetimes.

Table 4-10. Total Social Costs: Base case with zero discount rate and no new technology or research for new technology. Total costs are in $billions ($ × 10^9); lowest cost options are underlined. This table is a counterpart of Table 4-1.

Fuel Cost Escalation	Mix (fraction nuc) / Consumer Demand	4%	7%	Exponential to Log	Linear
0	0	2291	4540	2014	1780
	1/4	2194	4291	1932	1717
	1/2	2097	4043	1850	1654
	3/4	2001	3796	1768	1591
	1	_1905_	_3551_	_1686_	_1529_
Nuclear Fuel Escalation = 0.0003	0	2308	4557	2031	1797
	1/4	_2271_	_4550_	1997	1768
	1/2	2279	4831	_1993_	_1758_
	3/4	2330	5400	2021	1766
	1	2425	6256	2080	1792
Fossil Fuel Escalation =	0	3190	8682	2741	2339
	1/4	2837	6759	2479	2169
	1/2	2558	5315	2268	2030
	3/4	2353	4351	2109	1922
	1	_2220_	_3866_	_2001_	_1844_

Table 4-11. Total Social Costs: Base case with 5% discount rate and no new technology. Human costs are described in the counterpart of this table for the case of successful new technology, Table 4-2. Total costs are in $billions ($ × 10⁹); lowest cost options are underlined.

Fuel Cost Escalation	Mix (fraction nuc)	Consumer Demand	4%		7%		Exponential to Log		Linear	
0	0		1067	1225	1905	2240	984	1123	882	1000
	1/4		1032	1164	1816	2082	953	1070	859	960
	1/2		998	1103	1726	1925	923	1018	836	921
	3/4		962	1043	1638	1769	892	966	813	881
	1		_928_	982	_1549_	1615	_862_	915	_790_	841
Nuclear Fuel Escalation = 0.0003		0	1074		1912		991		889	
		1/4	1059		_1898_		976		877	
		1/2	_1056_		1970		_971_		_871_	
		3/4	1067		2128		976		_870_	
		1	1092		2373		990		875	

62 TOSCA

Table 4-11. continued

Fuel Cost Escalation	Mix (frac- tion nuc)	Consumer Demand			
		4%	7%	Exponential to Log	Linear
Fossil Fuel Escalation = 0.0003	0	1396	3202	1265	1109
	1/4	1283	2612	1177	1053
	1/2	1193	2164	1106	1006
	3/4	1125	1861	1052	969
	1	_1079_	_1700_	_1013_	_941_

Analysis 63

Table 4–12. Total Social Costs: Base Case with 10% discount rate and no new technology. This table is the counterpart of Table 4–3. Total costs are in $billions ($ × 10^9); lowest cost options are underlined.

Fuel Cost Escalation	Mix (fraction nuc)	Consumer Demand			
		4%	7%	Exponential to Log	Linear
0	0	606 / 822	975 / 1426	579 / 770	528 / 692
	1/4	591 / 773	938 / 1298	565 / 728	518 / 660
	1/2	577 / 724	901 / 1171	552 / 686	508 / 628
	3/4	562 / 675	864 / 1045	539 / 645	499 / 596
	1	_548_ / 626	_827_ / 920	_526_ / 603	_489_ / _564_
Nuclear Fuel Escalation = 0.0003	0	609	978	582	531
	1/4	601	_967_	574	526
	1/2	_598_	985	_571_	522
	3/4	_600_	1031	_570_	_520_
	1	606	1106	573	_520_

Table 4-12. continued

Fuel Cost Escalation	Mix (fraction nuc) / Consumer Demand	4%	7%	Exponential to Log	Linear
Fossil Fuel Escalation = 0.0003	0	749	1439	708	638
	1/4	709	1236	675	617
	1/2	676	1080	648	599
	3/4	650	972	626	585
	1	<u>632</u>	<u>912</u>	<u>610</u>	<u>573</u>

Table 4–13. Total Social Costs: Base case with a discount rate of 20% and no new technology; Table 4-4 gives the figures for the scenario of successful new technology. Total costs are in $billions ($ × 10⁹); lowest cost options are underlined.

Fuel Cost Escalation	Mix (fraction nuc)	4%	7%	Exponential to Log	Linear
0	0	298 / 553	413 / 932	294 / 522	277 / 472
	1/4	294 / 510	403 / 820	291 / 485	275 / 444
	1/2	291 / 467	394 / 709	287 / 449	272 / 417
	3/4	287 / 424	384 / 599	284 / 413	270 / 389
	1	<u>284</u> / 382	374 / <u>489</u>	<u>281</u> / 377	<u>267</u> / 361
Nuclear Fuel Escalation =	0	299	414	295	278
	1/4	297	<u>409</u>	293	276
	1/2	<u>295</u>	<u>408</u>	<u>291</u>	<u>275</u>
	3/4	<u>294</u>	411	<u>290</u>	<u>274</u>
	1	<u>293</u>	418	<u>289</u>	<u>273</u>

Table 4-13. continued

Fuel Cost Escalation	Mix (fraction nuc) / Consumer Demand	4%	7%	Exponential to Log	Linear
Fossil Fuel Escalation = 0.0003	0	343	505	338	318
	1/4	336	471	332	314
	1/2	330	444	326	310
	3/4	324	424	321	307
	1	<u>320</u>	<u>411</u>	<u>317</u>	<u>304</u>

The Contributions of Categories of Costs

It is helpful in interpreting the calculations to see how the various categories contribute to the total costs. Table 4–18 breaks into categories the costs from new plants for all-coal and all-nuclear scenarios when demand for electricity grows at 4 percent per year, the discount rate is 10 percent, and new technology is introduced.

In addition to costs considered previously, Table 4–18 includes in parentheses increments and totals arising from a term in human costs from cumulative effects of nuclear power. This cost is taken with an arbitrary coefficient of $1 million per reactor year; we have introduced it to show how a cumulative health effect would influence the total cost, not to represent any particular health effect.

Table 4–14. Incremental Cost Factors for Category 1 Costs and No New Technology. Table 4–5 is the counterpart of this table for successful new technology; the caption of Table 4–5 describes how to use these tables.

Discount Rate (%) \ Consumer Demand	4%	7%	Exponential to Log	Linear
0	98	272	68	58
5	42	111	32	26
10	21	54	17	14
20	7	20	7	5

Table 4–15. Incremental Cost Factors for Category 2a Costs and No New Technology. See Table 4–6 for a description of the use of this table.

Discount Rate (%) \ Consumer Demand	4%	7%	Exponential to Log	Linear
0	10.8	27.6	9.1	7.0
5	3.9	10.0	3.4	2.6
10	1.6	4.2	1.5	1.1
20	0.4	1.1	0.4	0.3

Analysis 69

Table 4—16. Incremental Cost Factors for Category 2b Costs with No New Technology. See Tables 4—6 and 4—7 for a description of how to use this table.

Discount Rate (%) \ Consumer Demand	4%	7%	Exponential to Log	Linear
0	22	100	17	12
5	7.3	32.2	5.9	4.1
10	2.8	11.4	2.3	1.6
20	0.60	2.2	0.50	0.40

Table 4—17. Incremental Cost Factors for Category 3 Factors with No New Technology. See Table 4—8 for a description of how to use this table.

Discount Rate (%) \ Consumer Demand	4%	7%	Exponential Log	Linear
0	75	157	71	59
5	26	54	25	21
10	11	21	10	9
20	2.6	5	2.3	2.3

Table 4-18. Contributing Costs of Various Kinds. The total 30-year costs, in $billions ($10^9$) from the most important categories are given for the new plants built to meet electrical demand growing at 4% per year, with the discount rate 10% and new technology being introduced. Human costs are discounted <u>except</u> where noted.

Scenario		Capital ($\sim\Delta N$)	Operating Costs ($\sim N$)	Nonescalating Part of Fuel Costs ($\sim N$)
All Coal	No escalation of coal cost	58.8	68.4	22.2
	No coal cost escalation; human costs not discounted	58.8	68.4	22.2
	Coal price escalating = 0.0003	58.8	68.4	22.2
All Nuclear	No escalation of nuclear fuel cost	58.9	39.7	14.8
	No fuel cost escalation; human costs not discounted	58.9	39.7	14.8
	Nuclear fuel price escalating = 0.0003	58.9	39.7	14.8

Table 4-18. continued

Human Costs ($\sim N$)	Escalation of Fuel Cost ($\sim \eta$)	Cumulative Part of Human Cost ($\sim \eta$) (assumed nuclear)	Total from New Plants
20.8	0	0	170.2
116.8	0	0	266.2
20.8	31.6	0	202.4
1.3	0	(10)	114.7 (124.7)
7.3	0	(65)	120.7 (165.7)
1.3	18.3	(10)	133.0 (143.0)

THE SUPPLY PROBLEM

At this point it is appropriate to test our demand scenarios against the current assessment of fuel supplies. The quantity of coal presents no problem;[27] and we do not address the question here of the rate at which it could be supplied.

The amount of uranium reserves and resources has been a subject of considerable controversy.[7,28] For purposes of this book, the significant question regarding supply is that of how much uranium is available at a price consistent with the operation of LWRs. There are two conflicting camps regarding the amount of uranium available at a price of $30 (1974 dollars) per pound of U_3O_8; the most widely bruited figures at present are 3.5 million tons[7] of proven and potential reserves, and about 1.4 to 1.7 million tons,[29,30,31] but the figures carry large uncertainties. The U.S. Department of Energy estimated, as of January 1978, that reserves and potential resources of uranium recoverable at $50 per pound are 2.42 million tons.[32] Estimates of the amounts available at prices up to $150 per pound of U_3O_8 are so uncertain as to be irrelevant here.

There are three purposes in asking how well potential demand can be matched by supply. First, uranium scarcity alone might assure that a significant proportion of the U.S. electricity over the next thirty years is generated by coal, either because of fear of true shortages or because the price of uranium will increase relative to that of coal. Our model of escalating nuclear fuel costs is roughly what might be projected if supply follows anything like Searl's projections.[31]

We have calculated the amount of electricity generated by each fuel for the thirty-year period for a variety of scenarios. From these figures, we estimated the amount of U_3O_8 required to supply the nuclear reactors of each scenario under two sets of assumptions. For one case, we assume no nuclear fuel reprocessing; for the other, we assume both uranium and plutonium are reprocessed. In both cases we assume that natural uranium is the primary fuel (rather than uranium-thorium or plutonium-thorium mixtures), that uranium is depleted to 0.25 percent, and the capacity factor of the plants is 80 percent. (The figure of 80 percent is intended to give a low-side estimate, relative to what current capacity factors would imply.) These models are presented by Pigford and Yang[33] and are discussed in the "Report to the American Physical Society by the group on nuclear fuel cycles and waste management."[34] We take them to be illustrative rather than definitive because the estimates of the average weight

Analysis 73

of U_3O_8 required to generate a unit of electrical energy span a wide range,[35] from about 190 to over 270 tons/gigawatt-year (electric). Pigford and Yang's estimates of the lifetime fuel requirements for the two cases are, respectively, 6,970 short tons (6,335 metric tonnes) and 4,750 short tons (4,320 metric tonnes), per 1000 Mw_e plant, corresponding to 232 and 158 tons of U_3O_8 per gigawatt-year (electric).

Table 4-19 gives the amounts of electric energy supplied by each fuel over the thirty-year period for the all-coal and all-nuclear op-

Table 4-19. Electric Energy Generated by Fuel (in Gigawatt-years).

Growth Rate	Fuel Used	Coal	Uranium	Alternate	Total
4%	All nuclear, to new technology	1677	5826	2105	
	All coal, to new technology	6053	1451	2105	9608
	All nuclear	1677	7932	0	
	All coal	8158	1451	0	
7%	All nuclear, to new technology	1677	12080	5953	
	All coal, to new technology	12300	1451	5953	19710
	All nuclear	1677	18030	0	
	All coal	18260	1451	0	
Exponential to Logarithmic	All nuclear, to new technology	1677	5579	1333	
	All coal, to new technology	5805	1451	1333	8589
	All nuclear	1677	6912	0	
	All coal	7138	1451	0	
Linear	All nuclear to new technology	1677	4480	1171	
	All coal, to new technology	4706	1451	1171	7328
	All nuclear	1677	5651	0	
	All coal	5877	1451	0	

tions with and without the introduction of a new technology. Note that even with a decision to build no new plants of one type, there is a small residual demand for that fuel in order to supply plants already in existence.

We come to the crux of the supply question with the data in Table 4–20, the tonnages of U_3O_8 required to supply the nuclear reactor system in each of the thirty-two cases shown.

Table 4–20. Amounts of U_3O_8 to Fuel the Nuclear Power System (millions of short tons).

Growth Rate	Scenario	U_3O_8 Required without Reprocessing	U_3O_8 Required with Reprocessing of Uranium and Plutonium
4%	All nuclear, to new technology	1.35	0.92
	All coal, to new technology	0.34	0.23
	All nuclear	1.84	1.26
	All coal	0.34	0.23
7%	All nuclear, to new technology	2.81	1.92
	All coal, to new technology	0.34	0.23
	All nuclear	4.19	2.87
	All coal	0.34	0.23
Exponential to Logarithmic	All nuclear, to new technology	1.30	0.89
	All coal to new technology	0.34	0.23
	All nuclear	1.60	1.10
	All coal	0.34	0.23
Linear	All nuclear, to new technology	1.04	0.71
	All coal, to new technology	0.34	0.23
	All nuclear	1.31	0.90
	All coal	0.34	0.23

The sixteen "all-coal" scenarios have little relevance in this part of our discussion because they present no problem for suppliers. We have not included any intermediate cases here because they can be interpolated directly from the all-coal and all-nuclear cases. The all-nuclear cases span the range from a low requirement for U_3O_8 of 0.7 million ton, in the linear-electricity demand case with reprocessing of uranium and plutonium and with new (non-nuclear) technology, to a high requirement of 4.2 million tons with no reprocessing, a 7 percent growth of demand for electricity, and no reprocessing. An "intermediate" planning scenario might assume a 4 percent growth of demand, with no reprocessing and with or without a new technology. Without new technology, we find a uranium requirement of 1.8 million tons, the highest requirement apart from those scenarios based on 7 percent growth of demand. This figure is above the pessimistic estimates but well below the optimistic estimates of what is available in the ground at a forward price of $30 per pound. With new technology, the lifetime requirement of the system drops to 1.35 million tons, just attainable with the most pessimistic estimates.

If the real system uses a mixture of coal-fired and nuclear plants with least total cost choices made on a regional basis, then it is likely that we have the supply of uranium required for the next thirty years. However, we must keep in mind that this thirty years—or perhaps it will be fifty or sixty years, depending on how many nuclear plants there are—is the one brief interval in which we can fuel our generators with low-cost uranium. Thereafter, we shall have to find something else. These figures, of course, were part of the basis for choosing a thirty-year period for this model; it is roughly the longest time for which we could seriously consider an all-nuclear system consisting of LWRs as we have them. In order to have the option of going to a largely nuclear program for any longer period, we may want to adopt a demand pattern something like those described by the Supply Panel of the Committee on Nuclear and Alternative Energy Sources (CONAES).[36]

We have not probed the supply problem in greater depth. For example, matching the processing and enrichment plant capacity to the demand has not been considered. But we hope this excursion will encourage comparisons of costs and options on a regional basis so that we can make better estimates of what the real demand for uranium would be nationally if each power company were to choose its own lowest total cost mix.

EXTENDING THE CALCULATIONS

The content of this book is not only the evaluation of costs for specific mixes of nuclear and coal-fired power stations. At least as important, we hope, will be the provision of a simple technique for evaluating the total social cost of power plant mixes with quite different assessments of the contributing costs. The intent here is to give the reader a means to do approximate evaluations quickly and easily to determine the total social costs and least cost mixture from individual cost figures plausible for particular regions. As the following description and example will show, Tables 4–1 to 4–17 are the data base from which these calculations can be made by hand in a way that allows one to include virtually as many changes as one wishes.

To illustrate, we present two examples. First, we evaluate the total costs for all the mixes of a hypothetical scenario; second, we revise the 1974 estimates for the Chicago Commonwealth Edison case, updating them with 1978 costs.

A Hypothetical Western Region

This model might be imagined to represent a Western region where low-sulfur coal is readily available, prior to requirements for scrubbers, where the population density is low, and where the price of coal is low. The costs are all chosen arbitrarily but not without design. We consider that the choice of the mix may or may not incorporate risk aversion toward nuclear accidents, but it has no other risk aversion.

Let us take construction costs K_c to be \$320/kw (which could correspond to plants using low-sulfur coal), the coal cost E_c to be \$37.6 million/plant-year, and the human cost H_c to be \$6 million/plant-year. Furthermore, suppose that the risk aversion factor is either 1 (no risk aversion) or 30 and applies only to nuclear power, that fuel costs are constant, that demand grows 4 percent annually, that the discount rate is 10 percent, and that new technology is introduced. We therefore turn to the upper left of Table 4–3 to find base case costs as follows:

Fraction Nuclear	Base Costs (Billion $)
0	609
1/4	598
1/2	586
3/4	575
1	563

Let us compute the increments to total cost due to each change in factor cost. To evaluate these, we need the coefficients of Tables 4–5 to 4–9, which become multipliers of the changes in factor costs. Each one of these tables describes a different category of costs among the three categories described in Chapter 3, on contributing costs, and in Tables 3–1 and 3–2. Thus, the base case capital cost of coal-fired plants is $420/kw (Table 3–2); the cost assumed now is $320/kw so the change is −$100/kw. This cost contributes to the total cost in proportion to ΔN (Category 1), the number of new plants built. Therefore we turn to Table 4–5 and look for the incremental cost for a discount rate of 10 percent and growth rate of 4 percent. The coefficient is 14 and the unit of increment is −$100/kw so that the increment to total cost due to the lower capital cost is −$14 billion.

Next we evaluate the change in cost of fossil fuel. The base case cost (Table 3–2) is $52.6 million, the new cost is $37.6 million, and the change is −$15 million. The fuel costs are assumed constant so we use Table 4–6 rather than Table 4–7 (Category 2a rather than Category 2b) to provide the multiplier. For 4 percent growth and a 10 percent discount rate, the multiplier is 1.3. Therefore the change in fuel cost is (−$15)(1.3) = −$19 billion.

The decrease in human costs is $16 million − $6 million or $10 million. This cost may be taken proportional to N_c or to η_c. If we choose the former, then the decrease in human cost of $10 million per year should be multiplied by 1.3, as found in Table 4–6, giving a total reduction of $13 billion. Instead, if we suppose the health effects are cumulative, Table 4–8 should be used (Category 3 cost). Then the coefficient is 10, and the total cost is $100 billion. For illustration only, we choose the former and use −$13 billion. All the changes in costs discussed thus far pertain to coal-fired plants.

The only change we introduce here for a nuclear system is an increase in risk aversion from the risk neutral posture. If the risk aversion factor is 30, then the change in cost for nuclear plants is a change in human cost from $1 million (base case) to $30 million. The change of $29 million is multiplied by the coefficient 1.3 from Table 4–6 to give the total increase in cost of $38 billion for an all-nuclear scenario.

Now we evaluate the total costs of the five mixes, from all coal to all nuclear:

(See table overleaf)

Increment to Each Category

Fraction Nuclear	Base Cost	K_c	E_c	H_c	H_n without Risk Aversion	H_n with Risk Aversion	Total without Risk Aversion	Total with Risk Aversion
0	$609 billion	−14	−19	−13	+0	+0	563	<u>563</u>
1/4	598	$-14 \times \frac{3}{4}$	$-19 \times \frac{3}{4}$	$-13 \times \frac{3}{4}$	+0	$+38 \times \frac{1}{4}$	563	573
1/2	586	$-14 \times \frac{1}{2}$	$-19 \times \frac{1}{2}$	$-13 \times \frac{1}{2}$	+0	$+38 \times \frac{1}{2}$	563	582
3/4	575	$-14 \times \frac{1}{4}$	$-19 \times \frac{1}{4}$	$-13 \times \frac{1}{4}$	+0	$+38 \times \frac{3}{4}$	563	592
1	<u>563</u>	−0	−0	−0	+0	+38	563	601

Recall that the base cases include the nonrisk-averse contributions to H_n.

Analysis

Thus, without risk aversion, all five cases have the same cost; the hypothetical example was constructed to yield this result. With risk aversion, the lowest total cost occurs with an all-coal system.

To use examples with fuel costs escalating with a multiplier of 0.0003, one employs Table 4-7. For a different escalation rate, one must interpolate or extrapolate. Suppose that coal costs are constant and nuclear fuel costs escalate with a coefficient of 0.0001. We construct new base totals from Table 4-3.

Fraction Nuclear	Base with Zero Fuel Cost Escalation	Base Fuel Escalator $\beta = 0.0003$	Interpolated Base $\beta = 0.0001$
0	609	612	610
1/4	598	606	601
1/2	586	601	591
3/4	575	599	583
1	<u>563</u>	<u>600</u>	<u>575</u>

Subtract the base case with no fuel cost escalation from the interpolated base case with escalation and add the difference to the column, "Total without Risk Aversion" from the previous table. Now the totals without risk aversion become

Fraction Nuclear	Total Cost
0	<u>564</u>
1/4	566
1/2	568
3/4	571
1	575

The minimum cost is again with the all-fossil system.

Updating the Chicago-Commonwealth Edison Case for 1978 Cost Estimates

The base case calculations of Tables 4-1 to 4-4 and 4-10 to 4-13 were constructed with direct costs for the Commonwealth Edison Company in 1974. The costs of capital, operation and maintenance, and fuel have all increased since 1974. We now construct an updated set of total costs for 1978 for one of the scenarios using the 1978 direct costs[37] and the increment tables.

We take for our example a growth rate of electrical demand of 4 percent and a discount rate of 10 percent.

The cost adjustments can be considered one category at a time. First are the capital costs, proportional to the incremental number of new plants, ΔN. These costs contribute as shown in Table 4-21, based on Tables 4-5 and 4-14:

Table 4-21. 1978 Increments to Capital Costs.

	1978 Capital Cost ($/kw)	Change from Base Case	Increment to Total Cost ($billion)	
			With New Technology	Without New Technology
Coal-fueled	638	+218	2.18 × 14 = $30.52	2.18 × 21 = $45.78
Nuclear	692	+271	2.71 × 14 = $37.94	2.71 × 21 = $56.91

Table 4-22. 1978 Increments to Operating Costs.

	1978 Operating Cost (mills/kwh)	1978 Operating Cost ($million/plant-year)	Change ($million)	Increment to Total Cost ($billion)	
				With New Technology	Without New Technology
Coal-fired (large)	3.0	$9.72	−7.4	−7.4 × 1.3 = −9.62	−7.4 × 1.6 = −11.89
Nuclear (system average)	2.2	$11.57	+0.2	0.2 × 1.3 = 0.26	0.2 × 1.6 = 0.32

Note that the costs in Table 4-21 and in those that follow are evaluated with the estimated costs of new technology equal to those used for Tables 4-1 to 4-4. This assumption has the effect of raising the cost of any scenario using no new technology relative to the comparable case with new technology being introduced successfully.

The costs of operation and maintenance depend on N, the number of plants in service. In our model, we take these to be independent of capacity factor although parts of the operating costs may depend somewhat on L. The operating cost of Commonwealth Edison's large coal-fired plants dropped from the 1974 value of $17.1 million per plant year to 9.72 million, which may be due in part to a drop in capacity factor from 60 percent to 46 percent. The influence of operating costs, Table 4-22, is derived from figures in Tables 4-6 and 4-15.

Fuel costs are slightly more complex because they may increase with fuel used to date. These figures are based on capacity factors of 46 percent and 60 percent, respectively, for fossil and nuclear power plants, the 1977 figures for Commonwealth Edison. The coal figures are based on low-cost, high-sulfur coal being used in plants equipped with scrubbers adequate to meet Environmental Protection Act standards. (The costs of scrubbers are incorporated in the capital costs.) The costs of nuclear fuel include 2 3/4 mill/kwh for reprocessing waste disposal and carrying. Table 4-23 shows how the new fuel costs affect total costs (in $billion or 10^9). Thus when we use Tables 4-6, 4-7, 4-15 and 4-16, we obtain the increment given in that table.

The increments to the total costs due to the changes in direct costs between 1974 and 1978 can now be evaluated, as shown in Table 4-24. Here we consider separately systems based entirely on coal-fired or nuclear plants, with and without new technology and with fuel escalators β of zero and 0.0003, in $billion ($10^9$).

Table 4-23. 1978 Increments to Fuel Costs.

	Replacement Cost (mills/kwh)	Cost ($million/plant-year)	Change ($million/plant-year)	Increment to Total Cost ($billion)			
				With New Technology		With No New Technology	
				No Escalation	$\beta = 0.0003$	No Escalation	$\beta = 0.0003$
Coal-fired	12	63.12	10.52	10.52×1.3 = 13.68	10.52×1.9 = 19.99	10.52×1.6 = 16.83	10.52×2.8 = 29.46
Nuclear	7	35.56	5.06	5.06×1.3 = 6.58	5.06×1.9 = 9.61	5.06×1.6 = 8.10	5.06×2.8 = 14.17

Table 4-24. Sum of Increments to Total Costs, Based on 1974 to 1978 Changes in Direct Costs for the Commonwealth Edison Company.

	With New Technology		Without New Technology	
	$\beta = 0$	$\beta = 0.0003$	$\beta = 0$	$\beta = 0.0003$
Coal-fired	$30.52 - 9.62 + 13.68$ = 34.58	$30.52 - 9.62 + 19.99$ = 40.89	$45.78 - 11.84 + 16.83$ = 50.77	$45.78 - 11.84 + 29.46$ = 63.40
Nuclear	$37.94 + 0.26 + 6.58$ = 44.78	$37.94 + 0.26 + 9.61$ = 47.81	$56.91 + 0.33 + 8.10$ = 65.33	$56.91 + 0.32 + 14.17$ = 71.40

We can add these increments to the totals in Table 4−3 (for a successful transition to new technology) and Table 4−12 to give total social costs based on updated direct costs. Note again that these calculations are based on a growth of demand of 4 percent per year and a discount rate of 10 percent with human life discounted. These results are given in Table 4−25. The effects of the change are to shift the lowest cost mix toward more use of coal and less of nuclear, largely because of differential increases in capital costs, and to give an apparent advantage to the introduction of new technology. This effect is a consequence of our assumption that costs of new technology are not increasing. We have not tried to develop new estimates of these costs. An appropriate application of TOSCA would be the adjustment of Table 4−24 for reestimated costs of new technology. One might rationalize the constant cost assumption on the basis of greater learning and improved technology since 1974, but the differences between the first two columns of costs and their counterparts in the last two columns in Table 4−24 must be considered as the outcome of a speculative assumption.

The breakdown of the new direct costs by category, analogous to the breakdown of Table 4−18, is given in Table 4−26. The figures are all the present value of total cost in $billion ($10^9$) for each category for the total increment due to the increases in direct costs and, in the last column, the total, which is the sum of the total cost of new plants from Table 4−18 and the corresponding increment. These are all computed with 4 percent growth of demand, 10 percent discount rate, new technology, and human costs discounted.

Table 4-25. Total Costs, 1974 and 1978, Based on Direct Costs for the Commonwealth Edison Company (10^9).

Fuel Cost Escalation	Mix (fraction nuclear)	New Technology 1974	New Technology 1978	No New Technology 1974	No New Technology 1978
0	0	609	644	606	657
	1/4	598	635	591	645
	1/2	586	626	577	635
	3/4	575	617	562	624
	1	_563_	_608_	_548_	_613_
Nuclear Fuel Escalation = 0.0003	0	612	647	609	660
	1/4	606	_644_	601	_657_
	1/2	_601_	_642_	_598_	_659_
	3/4	_599_	_643_	600	666
	1	_600_	648	606	677
Fossil Fuel Escalation = 0.0003	0	725	766	749	812
	1/4	700	742	709	773
	1/2	698	741	676	740
	3/4	661	705	650	715
	1	_648_	_693_	_632_	_697_

Table 4-26. Breakdown by Category of Cost Increments (1978 costs—1974 costs) for New Plants.

	Capital ($\sim\Delta N$)	Operating ($\sim N$)	Fuel ($\sim N$)	Escalation of Fuel ($\sim \eta$)	Total Increment	Total for New Plants
All Coal, $\beta = 0$	30.5	−9.62	13.68	0	34.6	204.8
All Coal, $\beta = 0.0003$	30.5	−9.62	13.68	6.31	40.9	243.3
All Nuclear, $\beta = 0$	37.9	0.32	6.58	0	44.8	159.5
All Nuclear, $\beta = 0.0003$	37.9	0.32	6.58	3.03	47.8	180.8

Chapter 5

Concluding Viewpoints

We carefully avoid titling this final discussion "Conclusions," because there are never conclusions to the kind of question we posed at the outset. There is no permanent answer to the optimal mix of power plants in the way a formal mathematical problem has an answer. Our conclusions concern the nature of the answers to our question. TOSCA tells us that the optimal mix of fossil and nuclear plants is a sensitive function of capital and fuel costs, that it depends on ethical judgments such as whether or not to discount human life, and that it is insensitive to such factors as safeguarding against diversion and sabotage and to climatic considerations provided our presumptions stand the test of ongoing scientific investigations. Our conclusions tell us that the optimal mix varies from one locality to another and from one power company's estimate of its costs to another.

The calculations show that when we include all the inputs to which we can ascribe costs, there is often rather little difference among the total costs of alternative mixes. This is what one would anticipate for truly competitive technologies. TOSCA tells us that a sharp change in the cost of a single component is likely to make one technology much more attractive than the other. Moreover, we must recognize that factors to which we cannot ascribe costs may sometimes play the swing role for the choice of electrical supply systems.

Another implication of the TOSCA calculations bears on the reduction of uncertainties. The figures in Tables 4–1 to 4–17 let us compute the total cost from assumed factor costs. For example, if we assume a 5 percent discount rate, new technology, a 4 percent growth, and an escalating price of uranium, we select a mix between 50 and 75 percent nuclear (Table 4–2 in the 4 percent growth column). The total cost for this optimum is $1,049 \times 10^9$. Suppose that the cost of uranium does not rise but the cost of coal does, at the rate indicated in Table 4–2. Then the optimum mix would be all nuclear, with a cost of $1,112 \times 10^9$. Selecting a mix of 75 percent nuclear and 25 percent coal plants based on the assumption of rising uranium prices and constant coal prices would generate a total cost of $1,143 \times 10^9$, or $31 billion more than the optimum for a world with rising coal prices. This $31 billion is a measure of the value of reducing the uncertainty in future fuel costs. It is an upper bound to how much it is worth spending on mineral exploration and stockpiling to improve the likelihood that the path we select as optimal corresponds more or less to the way things happen.

Even when the calculations indicate that a utility should choose a single kind of power system, the utility may wish to incur the extra costs of a mixed system in order to insure its customers of a stable supply of electricity. The cost of diversity is a kind of insurance cost, equal to the difference between the cost of a single-fuel system and the chosen mixed system.

Whatever we do now about constructing power plants constitutes a policy choice. If we were to decide to build no new plants at all for some years, we would be making tacit assumptions that the present mix of fossil and nuclear plants would be the optimal way to supply electricity to a society whose demands could be met during those years by existing plants and that the social costs of replacing aging plants would outweigh the cost of operating those plants. Admittedly, our model does not include a constant or decreasing demand pattern, and it does not examine the alternative of extending lives of plants now in use. (Note that it would be consistent with our model to allow for future plants with useful working lives greater than thirty years, but this change would only affect models with longer time spans than ours.) Within the range of possible worlds we consider, one can take our figures or others derived from them as best present guesses, guidelines for how to behave with today's projections. The best way to use the TOSCA model would obviously be to reevaluate the factor costs annually and revise the total cost tables for each new updated set. The model then becomes a rolling model, hopefully with steadily diminishing uncertainties and increasing reliability. As new technologies come closer to reality, this kind of updating should become a more and more useful guide to tell us where and how rapidly to plan to introduce them.

The TOSCA analysis can only be expected to be one tool in the complex political and social process of a society selecting a technology. Hopefully, it will sharpen our perceptions and lead us to spend our efforts and our concerns on the most important issues.

Appendixes

✳ *Appendix 1*

Cost Estimates

DIRECT FINANCIAL COSTS

Facility Construction Costs, K

Coal and Nuclear Plants. In this investigation, both coal and nuclear plants are assumed to be 1,000 Mw installed capacity facilities, and scale economies are not assessed. The most comprehensive sources for data on current construction costs are the Form 1 Annual Reports of the Federal Power Commission.[38] Costs for both types of plants can be expected to be sensitive to geographical location, and we consequently have chosen an array of utility estimates and displayed these in the figures. Rather than an arbitrary construct based on averaged costs, this model uses for its base case the figures provided by Commonwealth Edison, which serves the Chicago and Northern Illinois area. This had the advantage that the figures were easily obtained and could be verified. We should point out again that the choice of this or any other base case does not affect the final figures one would compute for any other example. All data are in 1974 constant dollars.

Commonwealth Edison reports that construction of light water reactors and coal-fired plants with scrubbers costs approximately $420/kw.[39] Coal plants without scrubbers are considerably cheaper to build at $295/kw but EPA regulations prior to the 1977 Clean Air Act Amendments required that they burn only low-sulfur coal.[39] Philadelphia Electric projects their LWRs as more costly ($365/kw) than coal-burning generation stations ($234/kw with scrubbers and $175/kw without).[40] The estimates by Duke Power Company, serving North and South Carolina, are suspiciously low: $182/kw for nuclear construction in 1973–1974 and $162/kw for coal plants built in 1973–1975.[38] Other estimates for LWRs range from $550/kw to $836/kw, while coal facilities vary from $565/kw to $935/kw with scrubbers and up to $510/kw without scrubbers.[18,41–45] Because of the extremely large variation in published costs (LWR: $182/kw to $863/kw; coal with scrubbers: $162/kw to $935/kw; and coal without scrubbers: $175/kw to $565/kw) it is imperative that the sensitivity of the results to these costs be carefully tested. Note that we have not imputed an increase in the costs over time for either reactors or coal-fired facilities, other than those inflationary costs that are normalized out in a constant dollar calculation. By using a constant cost of capital, one is using an averaged cost similar to a levelized cost.[46] Estimates of decommissioning costs of a nuclear reactor are $3–45 million;[60] we used $24 × 10^6. This analysis is insensitive to the choice because only a few reactors go out of service during the 30-year period.

New Technologies. Costs for new technologies for which we have no experience must be projected from engineering designs. Concomitantly, the uncertainties are great. Breeder reactor costs have been estimated to be between $500/kw and $1,000/kw.[41,42] One source[42] projects breeder costs as 42 percent greater than a conventional LWR. For an LWR figure of $421/kw, the cost reported by Commonwealth Edison, the corresponding breeder cost investment would be approximately $600/kw, and we utilize this value in the base case calculation.

A recent California study[44] of solar central station construction costs provides a range of estimates from $750/kw to $2,000/kw. A Jet Propulsion Laboratory assessment[47] of orbital and terrestrial central power stations found that a large solar central steam plant employing a flat heliostat collector would incur a capital cost of $900/kw, while other solar technologies appeared feasible between $1,150/kw and $2,500/kw. We assume a base case value of $1,000/kw, assuming that the lowest capital cost technologies will achieve the greatest market penetration in this highly capital-intensive industry.

If TOSCA is used as a tool for continuing analyses, new technologies such as fluidized-bed combustion could be incorporated as one can foresee their deployment.

Operation and Maintenance Costs, L

As with construction costs, considerable variation is reported in operating costs (again in 1974 dollars). Philadelphia Electric cites costs of 9.2×10^6/yr for an LWR, 9.3×10^6/yr for a coal-fired plant without scrubbers, and 8.3×10^6/yr *additionally* for scrubber operation.[40] Corresponding Commonwealth Edison figures, which we again incorporate in our base case calculations for reasons given above, are 11.4×10^6/yr for both LWRs and coal facilities without scrubbers and a 5.7×10^6/yr add-on for scrubbers.[39] Estimates for a California location are 6.2×10^6/yr for LWR operation and maintenance and 13×10^6 for a plant that burns high-sulfur coal with scrubbing.[44]

Operating costs for new technologies are, of course, highly conjectural. A detailed study by Argonne National Laboratory estimates breeder operating costs to be 10 percent greater than a burner reactor,[48] and a figure of 12.5×10^6/yr is entered in our calculations. Although solar technology is popularly conceived as a low-cost technology; once in place washing and other maintenance of solar panels will require some effort in the form of labor. We estimate a cost of 40×10^6/yr, based on the cost of a staff of window washers required to maintain an equal area of ordinary windows.

Operation and maintenance expenses enter our calculation in the same way as nonescalating fuel costs ($\beta = 0$). In probing the sensitivity of the results to input factor costs, we have chosen a regime in which the fuel costs are varied over a wide range (see below). A reader who wishes to evaluate the impact of different operating and maintenance costs may examine the cases in which fuel costs are changed.

Fuel Costs, E

The basic fuel cost is the sum of the mineral mining, beneficiation, and transport costs. These costs are generally reported on a basis of per kwh generated (not per kwh delivered to final demand). Commonwealth Edison estimates that with uranium-containing yellowcake mineral selling at $30-35/lb, the nuclear fuel cost would be 5.5 mills/kwh.[39] The efficiencies of conversion of primary fuel energy to electric energy at the busbar are assumed to be 42 percent for coal and 33 percent for nuclear fuel.

For incorporation in our model, the per kwh figure must be converted to a yearly demand basis. The nuclear fuel cost is 28×10^6 per year for a 1,000 Mw plant, based on the actual average capacity factor of 58 percent for U.S. reactors.[39] New coal plants and plants with new technology are assumed to operate at a capacity factor of 60 percent. Like capital and fuel costs, these figures depend on the region and the utility, sometimes sensitively enough to affect the choice of mix.

The mix of high-sulfur Illinois coal and low-sulfur Western coal burned in scrubber-equipped facilities in operation in Illinois then costs 52.6×10^6 plant-year. Consumption of Western coal alone yields a cost of 84.2×10^6/plant-year but may not incur the capital and operating maintenance costs of scrubber operation. Other previously published estimates of uranium fuel costs vary from 15×10^6/plant-year to 25×10^6/plant-year and for high-sulfur coal, 49×10^6/plant-year to 70×10^6/plant-year.[18, 40, 41, 43, 48]

Among the new technologies, solar generating stations are assumed to use no fuel and to incur a zero fuel cost. Breeder reactors would be net producers of plutonium fuel eventually, but there would be fuel costs for the initial breeder reactor load. We estimate this cost to be 80 percent of that for an LWR, following the recent Argonne study.[48] This yields a figure of 23×10^6/plant-year. Another Atomic Energy Commission report evaluates breeder fuel cost at 4×10^6/plant-year,[41] but this seems unreasonably low over the time horizon we consider. A base case figure of 20×10^6/plant-year is arbitrarily adopted. The large and highly uncertain capital costs incurred dominate the decision as to adoption of breeder technology, and the calculation is relatively insensitive to assumptions regarding its fuel cost.

Finally, a projection of relative nuclear and coal fuel costs over time must be made. In all scenarios but one, costs of coal for plants on line before the program are taken constant, as if fixed by long-term contracts. The simplest assumption is that no change occurs for either fuel. Tables 4–1 to 4–4 and 4–10 to 4–13 are computed for this case, for a rapid increase in the cost of uranium only ($\beta_n = 3 \times 10^{-4}$) and for a like increase in the cost of coal only. One more complicated case was also computed. This case, which we call "base case," requires an increase in uranium prices, $\beta_n = 2 \times 10^{-4}$ in Equation 3–6. Coal costs rise only 10 percent as fast ($\beta_c = 2 \times 10^{-5}$). The nuclear cost rise would correspond to a doubling of the uranium price in twenty years if consumer demand grew at 4 percent and all new generation were nuclear. The faster rate of increase for nuclear fuel relative to coal is based on current estimates, which portray coal as relatively more abundant, relative to demand, than uranium ore.

In addition, one set of calculations assumed fuel prices increasing linearly with time with a doubling over the thirty-year period of both fuels including coal for preexisting fossil plants, while another set assumed a sudden jump (doubling) in year 8 of uranium or coal to reflect a price response to a sudden exogenous change in market conditions.

INDIRECT COSTS

Research R

Estimates of research can be derived from sources such as Energy Research and Development Administration (ERDA) and Department of Energy (DoE) budgets.[49] Research costs for current commercial plants are obtained by dividing that part of the budget attributable to research on LWRs or connected to coal combustion by the number of facilities in place. The result is 5×10^6 per plant (per year) for nuclear and 1×10^6 for coal. The cost of research for new technology is assumed to increase linearly with time from 1975 levels and at 1975 rates to a maximum that is treated as a variable. Once on line, a constant research cost for each unit is assessed, as for other operating plants.

For the expensive scenario modeled after breeders, the initial support level is 500×10^6 per year, increasing at 80×10^6/year to a maximum of 1060×10^6. Assumed research levels for operative breeder reactors are 5×10^6 per plant (identical to the LWR). Research costs for new solar power systems start at 15×10^6/year and rise 50×10^6/year to a maximum of 515×10^6.

Property Damage, PD

In the absence of strict environmental controls, substantial property damage is associated with the mining and combustion of coal. One source reports a range between 0.8×10^6/yr and 7×10^6/yr for pollution damage from power generation to property and materials.[24] Including effects from acid mine drainage, subsidence, and other mining-associated phenomena, an estimate of 30×10^6/plant-year has also been suggested.[50] However, both of these estimates were made prior to the Environmental Protection Act of 1969. Adherence to the provisions of this legislation should substantially reduce plant emissions and other effects. Thus, the suggested damage levels may be too high by a factor of ten. As a base line cost, we assume an environmental damage level of 3×10^6/plant-year, excluding damage to humans and the heat effects discussed below. In regions that are not arid, one study indicates that the cost of reclaiming the land used for coal mining is a small percentage of coal costs.[9] What reclamation costs will be in arid regions is not known. In either case, these costs will be incorporated in the market price of the coal under strict environmental regulations.

The principal environmental effects attributable to the nuclear fuel cycle are discussed in other sections. At present, so little uranium is mined that the environmental costs associated with its extraction are quite small when compared with those of coal mining.

Heat Loss and Regional Climate Change, T

There are two localized effects from heat release. First, heat release to the atmosphere can change the local climate (for cities, a heat island effect occurs), and the economic impact of this change may be positive or negative, depending on the region. Second, ecological changes occur due to heat released to bodies of water. We are interested primarily in the *difference* in these costs between coal and nuclear plants rather than their absolute magnitudes. This is because our model utilizes an exogenously determined demand for electricity that is not price-responsive. The sole difference in these costs therefore arises from the greater thermal efficiency of generation of coal-fired plants in comparison to LWRs.

The local temperature change per plant must be estimated. The temperature rise over a region associated with a net heat production and release to the atmosphere equivalent to 1 percent of insolation (heat received from the sun, ~ 250 watts/m^2) is about 2.5 degrees F.[12] Thus, a one-degree rise would result from 1 watt/m^2 of heat power released. For this calculation, we assume an efficiency of 42 percent for the newest coal plants and 33 percent for the corresponding nuclear facility. If the power plant is located near a city, so that the heat loss is spread over the urban area, a 1,000 Mw plant will raise the temperature 1 degree over an area of 3×10^9 m^2 (about 103 square miles) if it is a nuclear plant or 2.4×10^9 m^2 if it is coal-burning. The city of Chicago occupies 227 mi^2, so 1,000 Mw delivered to the metropolitan area by a nearby plant would raise the temperature about 0.5 degrees F. Present power generation in that area is somewhat greater than 2,000 Mw, so local changes no greater than one degree are expected. This calculation neglects heat removal by air circulation.

The change in fuel use per degree of temperature rise can be calculated from degree-day data and an equation that gives electricity use for heating and cooling as a function of temperature.[66] For Chicago, a one-degree increase in average temperature reduces total annual energy demand for direct heating and cooling. The one-degree rise would save 2 1/2 percent of the heating and cooling energy, 207 Gw-yr (thermal) or about 1.5×10^6 of coal annually. The maximum saving of 11 percent would occur for an eight-degree rise (if all heating is assumed to be by fossil fuels; the saving is more for electric heat). For a warmer climate, such as that of Cairo, Illinois, or Washington, D.C., any rise in temperature would increase energy use. If the economic impact is 1.5×10^6 per degree, the extra cost or saving per nuclear plant over the equivalent fossil plant, which releases 88 percent as much heat, is about 0.2×10^6. Thus this effect may be positive or negative, and the sign depends on the geographical region. In either case, it is so small that we have neglected it.

An estimate of the ecological cost of thermal pollution based on the consumptive use of water is taken to be 0.4×10^6 for coal plants and 0.6×10^6 for LWRs, so the differential of 0.2×10^6 is again small.[24] Even if the ecological consequences are ten times the above values, the calculation is relatively insensitive to the 2×10^6/yr difference.

A recent paper by Dyson[51] proffers an upper bound on the cost of global thermal effects. In addition to direct heat loss as discussed above, much attention has been given to increased CO_2 levels producing a global warming due to the so-called "greenhouse" effect.[12] This effect is associated solely with fossil fuel consumption. Reduction in CO_2 levels possibly could be accomplished by planting large numbers of trees, which would respire the CO_2, thereby lowering the global temperature. Dyson computes a cost of about 3.2×10^6 per year to plant enough trees to counteract the CO_2 thermal effects from a 1000 Mw fossil generating station. However, this calculation ignores the opportunity cost of planting trees instead of another, potentially more valuable crop on the production land. Noting the crude tradeoff between the coal plant's advantage in local thermal ecological impact and its potential damage through CO_2 production, we have neglected both costs in our model.

Another possibility not included here is interference by regional heating in the global circulation pattern. If the circulation were sufficiently unstable, the impact on agriculture and the attendant costs could be significant. This effect, like the impact of CO_2 on climate, is one whose costs should be included in future applications of TOSCA as they become more clearly understood.

SAFETY AND DAMAGE TO HUMAN HEALTH

Fuel Cycle and Operating Hazards (included in H)

This category of costs could equally well be labeled "endemic accidents and occupational hazards," events that occur sufficiently often that the frequency and consequences are approximately known and accepted as part of "normal" operation. Less likely events are treated under "accidents and contingencies."

Several estimates of human health costs[24,52-55] for light water reactors concur that these costs are small and within the range between 0.1×10^6 and $1. \times 10^6$ per reactor year. We take $1 million as the base case value and examine variations from this associated with risk aversion.

Similar estimates for coal-mining accidents and hazards in coal mining, transportation, and combustion range from $3.1 - 9.2 \times 10^6$ per reactor year.[18,53-60] This excludes effects of pollution from the combustion process. Much of the variance in these estimates is due to uncertainties concerning the effects of the Federal Mine Health and Safety Act. The estimate for these costs is included with the human costs of pollution (below) to give a total base human cost for coal plants of 16×10^6/plant-year. Obviously very long-term costs such as 100,000-year effects of radioactive tailings, if undiscounted, would overwhelm all other costs; such costs are omitted here.

Our highest estimates of damages expected for breeders and solar stations are small compared with the large but highly uncertain initial capital costs of these facilities. When discounted, the costs to new technology associated with accidents, 3×10^6/plant-year for breeders and 2×10^6 for solar, become a negligible part of the total social cost.

Pollution, H

Estimates of the effect of pollution from fossil-burning power plants on human health range from 1 to 110 excess deaths per plant-year. One study finds that an urban plant location is approximately three times as hazardous as a rural one.[57] Another estimates a twentyfold differential, which varies with the degree of enforcement of air pollution laws.[18] We take a cost of 16×10^6 per plant-year as the base value for pollution costs of coal plants and examine the sensitivity of our results to wide variations in this figure. In addition, we consider a scenario in which such hazards increase as more fuel has been burned. This has been accomplished by allowing β_{coal} to vary from 0 to 0.002.

Spent Fuel Storage

This factor contributes only to the cost of operating light water reactors. Two estimates are in good agreement on the one-time cost of reprocessing and storage of nuclear wastes without credit for Pu recovered. Commonwealth Edison estimates 0.5 mill/(kwh$_e$) or about 2.5×10^6/year,[9] while Allied General Nuclear Service cites

0.4×10^6 for treatment, 0.6×10^6 for transportation, and 1.8×10^6 for storage, totaling 2.8×10^6 per year.[61] A figure of 2.5×10^6/yr is used as a base case value.

There are two ways worth considering for the disposal of spent fuel, the current practice of perpetual care and one-shot disposal, which might become an option. If it is necessary to guard spent fuel in perpetuity, the cost each year is proportional to the total fuel used to that date. We can estimate the present cost of handling nuclear wastes as follows: 110×10^6 is the annual cost of handling military nuclear wastes, which total 1,000 times the volume of civilian wastes.[62] We divide this by the number of reactor years of civilian reactor service to give 0.003×10^6 per reactor year per year. Stringent storage requirements may be imposed in the future, and human hazards of storage must be added to the direct financial costs, so we increase the above figure by a factor of 100 to yield 0.3×10^6 per reactor year per year as the base estimate, again examining variations.

The second, presently more expensive alternative, a one-shot disposal, sets a crude upper bound on the cost of perpetual waste storage. It has been suggested that one could, in principle, recycle the accumulated actinides and blast the residual strontium and cesium into space at a cost of $6 to 17.5×10^6 per year's wastes.[63,64] If the conjectural figure of 17.5×10^6 is taken to be an upper bound to what should be spent on perpetual storage of one year's wastes, PS, then

$$\sum_{t=0} \frac{PS}{(1+r)^t} = \$17.5 \times 10^6, \text{ in which } r \text{ is the discount rate and } t \text{ is the year}$$

The solution to this simple relation gives values of PS as follows:

Discount Rate	PS
0%	0
10	1.5×10^6
20	2.5×10^6

Thus if the future is not discounted, the infinite cost of perpetual care guarantees that the proper course, *within the framework of the simple evaluation*, is disposal in space. As the discount rate goes up, so does the maximum one would pay for storage. Here, we do not claim that either the 2.5×10^6 or 17.5×10^6 figure is correct; they are used to illustrate the basic procedure for assessing the trade-off between the two possibilities. In the base case, we have used the 2.5×10^6 value.

ACCIDENTS AND CONTINGENCIES

Accidents, A

The Rasmussen report concludes that the expected number of deaths from reactor accidents per year of light-water operation is about 0.05.[27] The maximum within the error bars is 0.2 death expected per year. The figure of 0.05 was derived by multiplying together the number of deaths associated with accidents of different magnitudes and the probabilities of those accidents. This includes eventual deaths from long-term consequences such as thyroid cancers and genetic defects. The figure of 0.05 lives per reactor year is multiplied by \$1,000,000 per life, to obtain the cost of lives lost in accidents as \$50,000 per reactor year.

Included in the \$50,000 is the cost associated with a "worst plausible accident." These figures, from the Rasmussen Report, have been challenged as too low.[65] Let us suppose that the Rasmussen estimate is greatly in error, too low by a factor of 1,000, so that the probability of occurrence is one in 10^6 reactor years. Such an accident is estimated to cause approximately $\$14 \times 10^9$ damage to property and $\$50 \times 10^9$ to people, including effects that are slow to appear. This possible consequence is frightening and the sums involved huge. Yet multiplying the costs $\$64 \times 10^9$ times the arbitrarily increased probability of occurrence of 10^{-6} yields a cost of only $\$64 \times 10^3$/plant-year.

One might think that another way of examining risks of nuclear plant operation is to ask what is the cost of insurance against nuclear accidents. The Price-Anderson Act insures plants for damages up to a maximum of \$560 million.* This is clearly inadequate for a bad accident. We can ask the actuarial question, "What would the premium be for adequate coverage?" It is purely a theoretical question, since no insurance company has the reserves for the possible payoff. The annual premium would be the expected cost of an accident, cost × probability of occurrence. This equals the one-time cost of an accident, amortized over the period between such events. Hence we arrive at the same figure as before without an independent estimate.

Sabotage, TS

There are two estimates of safeguarding nuclear plants in the United States. Willrich and Taylor[20] consider "adequate" safeguards to cost less than \$1 million per reactor year for light water reactors and their support facilities, including recycling plants. An AEC re-

*Eventually \$1 billion.

port[41] gives the following capital and operating costs for safeguarding breeders and light water reactors, with and without plutonium recycle.

	LMFBR	LWR	LWR with Plutonium Recycle
Capital × 10^6	15.2	9.4	12
Operating × $ 10^6/yr	5.4	2.9	3.8

Thus, approximately $3 × 10^6 per plant year might be spent on effective antisabotage insurance for light water reactors and somewhat more for breeders. This is approximately equivalent to a 10 percent increase in fuel price. The initial capital costs are on the order of one ten-thousandth of the capital cost of the plant itself. The implication of these numbers is that the cost of effective safeguards is small compared to the alternative, whose cost should be estimated on the basis of the costs of a large accident.

 *Appendix 2*

The Program

The following pages are listings of the program used to compute the base case tables and all the other scenarios. The computations were carried out on a Data General Nova Model 1220 Minicomputer. Figure A2–1 is a flow diagram of the program.

Figure A2−1. Flow Chart.

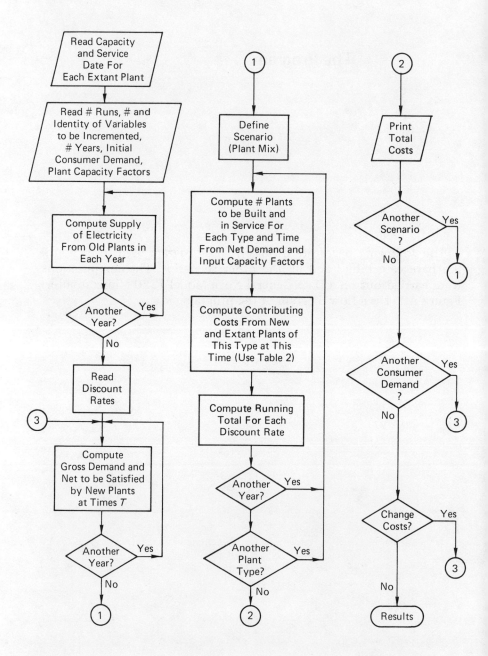

DESCRIPTION OF THE COMPUTER PROGRAM

Capacities and initial service dates for power plants in service in the United States in December 1974 are read by the computer. These data, with the assumed plant lifetime and capacity factors (contained in the program), are used to calculate the capacity in service and the electrical energy generated by that capacity in each year of the program. Data are read that specify the run. These include the number and identity of variables to be incremented, the form of consumer demand, the number of years, and the capacity factors for new plants. The consumer demand is calculated for each year, and from this total is subtracted the power generated by extant plants in that year, thus fixing the power to be supplied by new plants. The scenario is defined by the mix of plants to be built, including any new technology. Then the number of new plants of each type to go into service in each year ($\Delta N_i(t)$ in text) is calculated from the mix defined by the scenario and the consumer demand. The total number of plants of each type in service at each time, $N_i(t)$, is computed, as well as the total fuel of each type used to date, $\eta_i(t)$. These parameters are multiplied by the appropriate costs (base case costs from Table 3-2 or suitably incremented costs for any other scenario) and summed to give the cost in each year. Costs for each year are discounted at the chosen rate and a sum of total costs for each year computed. Total costs are printed, and the process is repeated for the next scenario. The entire procedure is repeated for a new consumer demand. Finally, one or more costs is incremented, and the procedure repeats from the first consumer demand and the first scenario.

```
LIST
25    DIM F[10],A[3,3]
28    DIM V[3,30]
50    DIM B[30],C[3,10],D[31]
75    DIM G[3,30],H[3],K[3]
100   DIM M[3,45],N[3,30],O[2,2],P[3,30],R[6]
150   DIM Y[3,30],Z[3,30]
160   DIM U[3,6],X[3,30],S[15,10],W[4]
200   DATA  1940,  1974,  4.54,  4.54,  4.54,  4.54,  4.54,  4.54,  4.54
210   DATA  4.54,  4.54,  4.54,  4.54,  7.6,  7.6,  7.6,  7.6,  7.6,  9.2
220   DATA  9.2,  9.2,  9.2,  9.2,  11,  11,  11,  11,  11,  13.5,  13.5,  13.5
225   DATA  13.5,  18,  16,  21,  22,  17
230   DATA  1957,  1984,  .09,  0,  .2,  .18,  0,  .34,  .06,  0,  .16,  0,  .9
235   DATA  .79,  .95,  1.74,  2.31,  4.71,  7.32,  12.54,  7.52,  8.45,  6.14
240   DATA  5.85,  13.56,  2.86,  6.72,  5.39,  .86,  .36
260   REM   0=OLD FOS,1=NUC,2=NEW FOS,3=NEW TECH
265   FOR J= 0 TO 3
270     FOR N1= 0 TO 45
275       LET M[J,N1]= 0
280     NEXT N1
285   NEXT J
300   FOR I= 0 TO 1
310     READ O[I,1],O[I,2]
315     LET O1=1985-O[I,1]
316     LET O2=1985-O[I,2]
320     FOR N1=O1 TO O2 STEP -1
330       READ M[I,N1]
340     NEXT N1
350   NEXT I
360   LET C8=1
380   LET C5= 0
397   PRINT "#PLANT TYPES-1,#CONSTANTS,#RUNS"
398   INPUT J1,C1,C9
400   PRINT "# VARIABLE INCREMENT RUNS";
405   INPUT I1
410   IF I1= 0 GOTO  460
415   FOR I2=1 TO I1
420     PRINT "#VAR. IN SET";I2;"FACTOR";
425     INPUT S[I2, 0],F[I2]
430     LET S1=S[I2, 0]
435     FOR S2=1 TO S1
440       PRINT "C( , )";
445       INPUT S[I2,S2],S[I2,S2+5]
450     NEXT S2
455   NEXT I2
460   FOR J= 0 TO 3
462     FOR B7=1 TO 3
465       LET A[J,B7]= 0
467     NEXT B7
470   NEXT J
472   LET A[1,2]=.0002
475   LET A[1,3]=.0005
477   LET A[ 0,2]=.0001
480   LET A[2,2]=.0001
482   LET A[ 0,3]=.0003
485   LET A[2,3]=.0003
```

Appendix 2 — The Program

```
487     PRINT "FUEL CST ESCALATIONS"
490     FOR J= 0 TO 3
492        PRINT A[J,1];A[J,2];A[J,3]
495     NEXT J
500     PRINT "TOTAL YEARS,BASE DEMAND";
550     INPUT T1,D0
551     FOR J= 0 TO J1
552        FOR T2= 0 TO T1
554           LET G[J,T2]= 0
555           LET B[T2]= 0
556        NEXT T2
560        PRINT "C.F.";J
561        PRINT
570        INPUT K[J]
585     NEXT J
600     LET T3=T1+1
620     FOR J= 0 TO 1
640        LET O1=O[J,1]
650        LET O2=O[J,2]
660        FOR N1=O1 TO O2
670           LET L5=N1-1975
700           LET L6=L5+34
820           IF L5>= 0 GOTO 860
840           LET L5= 0
860           IF L6<=T3 GOTO 900
880           LET L6=T3
900           FOR T2=L5 TO L6
905              LET G[J,T2]=G[J,T2]+M[J,1985-N1]
910              REM    COMPUTE SUPPLY FROM OLD PLANTS
920              LET B[T2]=B[T2]+K[J]*M[J,1985-N1]
930           NEXT T2
950        NEXT N1
960     NEXT J
970     PRINT "#DISC.RATES"
975     INPUT K9
980     PRINT "RATES"
985     FOR K2= 0 TO K9
986        INPUT R[K2]
987     NEXT K2
988     GOTO 8000
990     PRINT "COST CONSTANTS"
992     FOR J= 0 TO J1
993        FOR C2=1 TO C1
994           PRINT C[J,C2];
995        NEXT C2
996        PRINT
997     NEXT J
998     LET K1=.04
999     LET D1= 0
1000    FOR T9= 0 TO T3
1050       LET D[T9]=D0* EXP (K1*T9)-B[T9]-40
1100    NEXT T9
1110    PRINT "DISCOUNT RATES"
1112    FOR K2= 0 TO K9
1114       PRINT TAB (10*K2);R[K2];
1116    NEXT K2
```

```
1118      PRINT
1120      PRINT "CONSUMER DEMAND GROWTH";R1
1150      LET D[ 0]= 0
1698      LET P2= 0
1699      PRINT "ALL FOS TO NEW TECH"
1700      FOR T2= 0 TO T1
1710         LET E[ 0,T2]= 0
1720         LET E[ 1,T2]= 0
1730         LET E[ 2,T2]= 0
1740         LET P[ 0,T2]= 0
1750         LET P[ 1,T2]= 0
1770         IF T2>=T1/2 GOTO 1830
1790         LET P[ 2,T2]=1
1800         LET P[ 3,T2]= 0
1810         GOTO 1890
1830         LET P[ 2,T2]=1-T2/T1
1850         LET P[ 3,T2]=T2/T1
1890         GOTO 7350
1900      NEXT T2
1999      REM     COSTB- LOOP
2000      FOR J= 0 TO J1
2002         LET H[J]= 0
2005         LET W[J]= 0
2014      NEXT J
2015      FOR B7=1 TO 3
2025         FOR R2= 0 TO R9
2027            LET L[B7,R2]= 0
2030         NEXT R2
2035      NEXT B7
2040      FOR T2= 0 TO T1
2050         IF D[T2+1]<= 0 GOTO 2107
2070         IF D[T2]>= 0 GOTO 2101
2080         LET D2=D[T2+1]
2100         GOTO 2110
2101         LET D2=D[T2+1]-D[T2]
2105         IF D2>= 0 GOTO 2110
2107         LET D2= 0
2110         FOR J= 0 TO J1
2115            LET X[J,T2]= 0
2120            LET Y[J,T2]= 0
2125            LET H1=D2*P[J,T2]/K[J]
2127            LET H[J]=H[J]+H1
2130            LET N[J,T2]=H[J]+G[J,T2]
2140            IF N[J,T2]= 0 GOTO 2450
2150            LET W[J]=W[J]+N[J,T2]*K[J]
2160            IF T2>10 GOTO 2190
2165            LET N1=10-T2
2170            LET H2=M[J,N1]+H1
2180            GOTO 2200
2190            LET H2=H1
2200            LET X[J,T2]=X[J,T2]+C[J,1]*H2
2300            LET X[J,T2]=X[J,T2]+C[J,2]*N[J,T2]
2305            IF J=3 GOTO 2345
2310            IF J=1 GOTO 2345
2320            LET W2=W[ 0]+W[ 2]
2325            GOTO 2400
```

Appendix 2—The Program

```
2345        LET W2=W[J]
2400        LET Y[J,T2]=Y[J,12]+C[J,5]*N[J,12]*(1+C[J,6]*W2)
2450        LET Y[J,T2]=Y[J,12]+C[J,7]*W[J]+C[J,8]*N[J,12]
2460        IF J<>1 GOTO  2700
2470        LET X[1,12]=X[1,12]+24*M[1,45-T2]
2700        LET X1=X[J,12]+E[J,12]
2750        LET Y1=Y[J,12]
3000        FOR B7=1 TO 3
3050          LET V[B7,12]=N[J,12]*(1+A[J,B7]*W2)*C[J,3]
3100          FOR K2= 0 TO K9
3140            IF K2>3 GOTO  3180
3150            LET X3=X1+Y1+V[B7,12]
3160            GOTO  3200
3180            LET X3=X1+V[B7,12]
3200            LET S6=X3/((1+K[K2])↑12)
3250            IF K2<=3 GOTO  3400
3300            LET S6=S6+Y1
3400            LET U[B7,K2]=U[B7,K2]+S6
3420          NEXT K2
3450        NEXT B7
3470      NEXT J
3500    NEXT T2
3552    FOR B7=1 TO 3
3554      FOR K2= 0 TO K9
3556        PRINT  TAB (10*K2);U[B7,K2];
3560      NEXT K2
3561      PRINT
3562    NEXT B7
3563    IF C5<>1 GOTO  3616
3564    IF C8<>2 GOTO  3616
3565    LET C6=W[ 0]*.443+W[ 2]
3570    LET C7=N[2,11]+.443*N[ 0,11]
3575    PRINT "COAL";C6,C7
3580    PRINT "LWR";W[1];N[1,11]
3585    LET C6=.168*W[ 0]
3590    LET C7=.168*N[ 0,11]
3595    PRINT "OIL";C6;C7
3600    LET C6=.172*W[ 0]
3605    LET C7=.172*N[ 0,11]
3610    PRINT "GAS";C6;C7
3615    PRINT "ALTERNATE";W[3];N[3,11]
3616    PRINT
3620    LET P2=P2+1
3629    IF P2>11 GOTO  5000
3630    IF P2>1 GOTO  3900
3640    FOR T2= 0 TO 11
3650      LET P[2,T2]= 0
3700      IF T2>=T1/2 GOTO  3760
3720      LET P[1,T2]=1
3730      LET P[3,T2]= 0
3750      GOTO  3830
3760      LET P[1,T2]=1-12/11
3800      LET P[3,T2]=12/11
3830      GOTO  7350
3840    NEXT T2
3850    PRINT "ALL NUC TO NEW TECH"
```

```
3860     GOTO  2000
3900     IF P2>2 GOTO  4200
3950     FOR T2= 0 TO T1
3980        LET P[3,T2]= 0
4000        LET P[1,T2]=1
4020        LET P[3,T2]= 0
4050        LET P[2,T2]= 0
4100     NEXT T2
4130     PRINT "ALL NUCLEAR"
4150     GOTO  2000
4200     IF P2>3 GOTO  4500
4230     FOR T2= 0 TO T1
4250        LET P[1,T2]= 0
4280        LET P[2,T2]=1
4300     NEXT T2
4310     PRINT "ALL FOSSIL"
4350     GOTO  2000
4500     IF P2>4 GOTO  4620
4510     FOR T2= 0 TO T1
4520        LET P[1,T2]=.5
4530        LET P[2,T2]=.5
4550     NEXT T2
4580     PRINT "HALF AND HALF"
4600     GOTO  2000
4620     IF P2>5 GOTO  4800
4640     FOR T2= 0 TO T1
4660        LET P[1,T2]=.25
4680        LET P[2,T2]=.75
4690     NEXT T2
4700     PRINT "1/4 NUCLEAR, 3/4 FOSSIL"
4720     GOTO  2000
4800     IF P2>6 GOTO  6200
4810     FOR T2= 0 TO T1
4820        LET P[1,T2]=.75
4840        LET P[2,T2]=.25
4850     NEXT T2
4860     PRINT "3/4 NUCLEAR, 1/4 FOSSIL"
4880     GOTO  2000
4999     REM          CHANGE DEMAND
5000     LET D1=D1+1
5050     IF D1>3 GOTO  8000
5100     IF D1>1 GOTO  5250
5150     LET R1=.07
5200     GOTO  1000
5250     IF D1>2 GOTO  5500
5300     FOR T9= 0 TO 15
5320        LET D[T9]=D0* EXP (.04*T9)-B[T9]-40
5340     NEXT T9

5380     FOR T9=16 TO T3
5390        LET D[T9]=163.1* LOG (1.078*T9-5)-B[T9]-40
5395     NEXT T9
5400     PRINT "CONSUMER DEMAND EXP. TO  LOG"
5450     GOTO  1150
5500     FOR T9= 0 TO 13
5550        LET D[T9]=D0*(1+.04*T9)-B[T9]-40
```

Appendix 2 — The Program

```
5600    NEXT T9
5650    PRINT "CONSUMER DEMAND LINEAR"
5700    GOTO 1150
6000    STOP
6050    END
6200    IF P2>7 GOTO 6499
6250    PRINT "NUC,NEW TECH FAIL,COAL"
6300    FOR T2= 0 TO T1
6310        IF T2/T1>=.5 GOTO 6400
6350        LET P[1,T2]=1
6360        LET P[2,T2]= 0
6370        GOTO 6440
6400        LET P[1,T2]=1-T2/T1
6420        LET P[2,T2]=T2/T1
6440        GOTO 7350
6450    NEXT T2
6480    GOTO 2000
6499    IF P2>8 GOTO 6900
6500    PRINT "COAL WITH NEW TECH FAIL"
6550    FOR T2= 0 TO T1
6600        LET P[2,T2]=1
6650        LET P[1,T2]= 0
6700        GOTO 7350
6750    NEXT T2
6800    GOTO 2000
6900    IF P2>9 GOTO 9795
6902    PRINT "1/2 &1/2 TO NEW"
6904    FOR T2= 0 TO T1
6910        IF T2>=T1/2 GOTO 6990
6949        LET P[1,T2]=.5
6960        LET P[2,T2]=.5
6970        LET P[3,T2]= 0
6980        GOTO 7350
6990        LET P[2,T2]=.5-.5*T2/T1
7000        LET P[1,T2]=P[2,T2]
7100        LET P[3,T2]=T2/T1
7150        GOTO 7350
7200    NEXT T2
7250    GOTO 2000
7350    IF P[3,T2]= 0 GOTO 7500
7400    LET Z[3,T2]=(D[T2+1]-D[T2])*P[3,T2]*Z4/K[3]
7450    GOTO 7700
7500    IF T2>=(Z3-Z0)/Z1 GOTO 7650
7550    LET Z[3,T2]=Z0+Z1*T2
7600    GOTO 7700
7650    LET Z[3,T2]=Z3
7700    IF P2= 0 GOTO 1900
7750    IF P2=1 GOTO 3840
7800    IF P2=7 GOTO 6450
7850    IF P2=8 GOTO 6750
7860    IF P2=9 GOTO 7200
7870    IF P2=10 GOTO 9850
7880    IF P2=11 GOTO 9950
7900    STOP
8000    IF C5= 0 GOTO 8045
8002    IF C5=1 GOTO 8025
```

```
8004    LET S1=S[C6, 0]
8005    FOR S2=1 TO S1
8010      LET S3=S[C6,S2]
8015      LET S4=S[C6,S2+5]
8017      LET C[S3,S4]=C[S3,S4]/F[C6]
8020    NEXT S2
8025    LET C6=C5
8027    LET C5=C5+1
8028    IF C6>I1+2 GOTO 8045.
8029    IF C6<I1+1 GOTO 8032
8030    LET £3=£3*2
8031    GOTO 990
8032    LET S1=S[C6, 0]
8034    FOR S2=1 TO S1
8035      LET S3=S[C6,S2]
8037      LET S4=S[C6,S2+5]
8040      LET C[S3,S4]=C[S3,S4]*F[C6]
8041    NEXT S2
8043    GOTO 990
8045    LET C8=C8+1
8048    IF C8>C9+1 GOTO 6000
8050    FOR J= 0 TO J1
8070      FOR C2=1 TO C1
8090        READ C[J,C2]
8100      NEXT C2
8120    NEXT J
8130    IF C8<=2 GOTO 8140
8135    LET I1=1
8140    READ £0,£1,£3,£4
8148    LET C5=1
8150    GOTO 990
8500    DATA   295,  15,  75,  .0001,  16,  .000002,  0,  0
8510    DATA   421,  19.4,  30.5,  .0002,  0,  0,  .0003,  1
8520    DATA   420,  21.1,  52.6,  .0001,  16,  .000002,  0,  0
8530    DATA   600,  17.5,  25.5,  0,  0,  0,  0,  3
8560    DATA   500,  80,  1060,  5
8700    DATA   295,  15,  75,  .0001,  16,  .000002,  0,  0
8710    DATA   421,  19.4,  30.5,  .0002,  0,  0,  .0003,  1
8720    DATA   420,  21,  1,  52.6,  .0001,  16,  .000002,  0,  0
8730    DATA   1000,  40,  0,  0,  0,  0,  0,  2
8760    DATA   15,  50,  515,  2
9795    IF P2>10 GOTO 9900
9800    FOR T2= 0 TO T1
9805      IF T2>=T1/2 GOTO 9830
9810      LET P[1,T2]=.25
9815      LET P[2,T2]=.75
9825      GOTO 7350
9830      LET P[2,T2]=.75*(1-T2/T1)
9840      LET P[1,T2]=.25*(1-T2/T1)
9845      GOTO 7350
9850    NEXT T2
9855    PRINT "1/4 NUCLEAR TO NEW"
9860    GOTO 2000
9900    FOR T2= 0 TO T1
9905      IF T2>=T1/2 GOTO 9930
9910      LET P[1,T2]=.75
```

```
9915      LET P[2,T2]=.25
9925      GOTO  7350
9930      LET P[2,T2]=.25*(1-T2/T1)
9940      LET P[1,T2]=.75*(1-T2/T1)
9945      GOTO  7350
9950  NEXT T2
9955  PRINT "3/4 NUCLEAR TO NEW"
9960  GOTO  2000
```

References

1. A. Manne. "ETA: A Model for Energy Technology Assessment." *Bell J. Econ.* 7, 379 (1976).
2. W. Nordhaus. "The Allocation of Energy Resources." Brookings Papers on Economic Activity, No. 3 (1973).
3. Energy Research and Development Administration (ERDA). "The LMFBR—Its Need and Timing." ERDA–39 (May 1975).
4. K. Hoffman, and E. Cherniavsky. "A Model for the Analysis of Interfuel Substitutions and Technological Change." Upton, N.Y.: Brookhaven National Laboratory (1974).
5. W. Marcuse, L. Bodin, E. Cherniavsky, and Y. Sanborn. "A Dynamic Time-Dependent Model for the Analysis of Alternative Energy Policies." BNL–19406, Brookhaven National Laboratory (July 1975).
6. C. Komanoff. *Power Plant Performance. Nuclear and Coal Capacity Factors and Economics.* New York: Council on Economic Priorities, 1976.
7. S.M. Keeny, Jr., et al. *Nuclear Power Issues and Choices*, Report of the Nuclear Energy Policy Study Group. Cambridge, Mass.: Ballinger Publishing Co., 1977.
8. A.C. Pigou. *A Study in Public Finance*, 3rd ed. London: MacMillan and Co., 1947, Part I, Chapter V and Part II, Chapter IV.
9. T.V. Long. Study in preparation.
10. T.L. Brown. *Energy and the Environment.* Columbus, Ohio: Charles E. Merrill, 1971, p. 93.
11. H.H. Landsberg. *Man's Role in Changing the Face of the Earth.* W.L. Thomas, Jr., ed. Chicago: University of Chicago Press, 1956.
12. W.H. Mathews, W.W. Kellogg, and G.O. Robinson, eds. *Man's Impact on the Climate.* Cambridge, Mass.: MIT Press, 1971; (a) W.W. Kellogg, p. 126; (b) S. Manabe, p. 218; (c) W.M. Washington, p. 270ff.
13. E.J. Mishan. "Evaluation of Life and Limb: A Theoretical Approach." *Journal of Political Economy* 79, (1971): 687.

14, T.C. Schelling. "The Life You Save May Be Your Own." *Problems of Public Expenditure Analysis*. Washington, D.C.: The Brookings Institution, 1967.

15. S.E. Rhoads. "How Much Should We Spend to Save a Life?" *Public Interest* 51, 74 (Spring 1978).

16. M. Singer. "How to Reduce Risks Rationally," *Public Interest* 51, 93 (Spring 1978).

17. B. Cohen, "High Level Radioactive Waste from Light-Water Reactors," *Reviews of Modern Physics* 49, 1 (1977).

28. D.J. Rose, P.W. Walsh, and L.L. Leskovjan. "Nuclear Power—Compared to What," *American Scientist* 64, 291 (1976).

19. M. Willrich. "Terrorists Keep Out," *Bulletin of the Atomic Scientists* 31, 5, 12 (May 1975).

20. M. Willrich and T. Taylor. *Nuclear Thefts: Risks and Safeguards*. Cambridge, Mass.: Ballinger Publishing Company, 1974.

21. *Decision Making for Regulating Chemicals in the Environment*, Environmental Studies Board, Commission on Natural Resources, National Research Council (National Academy of Sciences, 1975).

22. U.S. Federal Power Commission. "The 1970 National Power Survey." Washington, D.C.: U.S. Government Printing Office, December 1971.

23. Federal Energy Administration. Project Independence Report. Washington, D.C.: U.S. Government Printing Office, November 1974.

24. "Nuclear Reactors Built, Being Built or Planned in the U.S. as of December 31, 1974." ERDA Office of Public Affairs, TIO−8200 (R31).

25. WASH−1224. "Comparative Risk-Cost-Benefit Study of Alternative Sources of Electrical Energy." (December 1975).

26. "Reactor Safety Study." U.S. Nuclear Regulatory Commission Report NUREG−75/014 (1975).

27. "Mineral Resources and the Environment." Report by the Committee on Mineral Resources and the Environment (COMRATE), Commission on Natural Resources, National Research Council (National Academy of Sciences, 1975).

28. U.S. Energy Research and Development Administration (ERDA), ERDA −76−1. "A National Plan for Energy Research, Development and Demonstration: Creating Energy Choices for the Future, 1976. Washington, D.C.: U.S. Government Printing Office, 1976.

29. M.A. Lieberman. "United States Uranium Resources—An Analysis of Historical Data." *Science* 192, 431 (1976).

30. J. Grynberg. Private communication.

31. M.F. Searl. "Uranium Resources to Meet Long-Term Uranium Requirements." EPRI Special Report No. 5 (November 1974); M.F. Searl and J. Platt, "Views on Uranium and Thorium Resources," *Annals of Nuclear Energy* 2, 751 (1975).

32. *Chemical and Engineering News* (June 5, 1978), p. 11.

33. T.H. Pigford and C.S. Yang. "Thorium Fuel Cycles." UCB−NE 3227, EPA 68−01−1962 (June 1977).

34. L.C. Hebel et al. "Report to the American Physical Society by the group on nuclear fuel cycles and waste management," *Reviews of Modern Physics* 50, 1, Part II (January 1978).

35. M.C. Day. "Nuclear Energy: A Second Round of Questions," *Bulletin of the Atomic Scientists* 31, 10, 52 (December 1975).
36. Demand and Conservation Panel of the Committee on Nuclear and Alternative Energy Sources. "U.S. Energy Demand: Some Low Energy Futures," *Science* 200, 142 (1978).
37. Commonwealth Edison Company. Summary of Costs (April 17, 1978).
38 Federal Power Commission, Form 1, Annual Reports.
39. A.D. Rossin. Commonwealth Edison, "Reliability and Economics of Light Water Reactors." American Nuclear Society, Public Information Committee White Paper (May 1976).
40. V.S. Boyer. Philadelphia Electric Co., "The Economics of Nuclear Power," Third Congressional Seminar on the Economic Viability of Nuclear Energy. Washington, D.C., June 7, 1976.
41. U.S. Atomic Energy Commission Report, WASH–1535. "Breeder Cost-Benefit Analysis." (December 1974).
42. J.J. Taylor. "The Status of the LMFBR Development, *Annals of Nuclear Energy* 2, 705 (1975).
43. "Economic Comparison of Base Load Generation Alternatives for the New England Electric System." Report to the Board of Trustees of Northeast Utilities by Arthur D. Little, Inc., and S.M. Stoller Corp., Cambridge, Mass. (March 1975).
44. "Impacts of Alternative Electricity Supply Systems for California." Western Interstate Nuclear Board, May 7, 1976.
45. H. Bethe. "The Necessity of Fission Power," *Scientific American* 234, 1, 21 (January 1976).
46. D.L. Phung. "Cost Comparison Between Base Load Coal-Fired and Nuclear Plants in the Midterm Future (1985–2015," Institute for Energy Analysis, Oak Ridge Associated Universities, Oak Ridge, Tenn., ORAV/IEA(M) 76–3 (September 1976).
47. L.D. Hamilton, ed. "The Health and Environmental Effects of Electricity Generation—A Preliminary Report," Upton, N.Y.: Brookhaven National Laboratory (1974).
48. Argonne National Laboratory. Report ANL 8092, February, 1973.
49. U.S. Energy Research and Development Administration. FY 1978 Budget Request. *The Energy Daily*, January 18, 1977.
50. W. Meyer (University of Missouri). 15th Annual Nuclear Engineering Education Conference, Argonne National Laboratory, March 8–9, 1976.
51. F.J. Dyson. Institute for Energy Analysis, Oak Ridge, Tenn., IEA–(0)–76–4 (1976).
52. L.A. Sagan. "Human Costs of Nuclear Power," *Science* 177, 487 (1972).
53. L.B. Lave and L.C. Freeburg. "Health Effects of Electricity Generation from Coal, Oil and Nuclear Fuel," *Nuclear Safety* 14, 409 (1973).
54. D.J. Rose. "Nuclear Eclectic Power," *Science* 184, 351 (1974).
55. B.L. Cohen. "Impacts of the Nuclear Energy Industry on Human Health and Safety," *American Scientist* 64, 550 (1976).
56. C.L. Comar and L.A. Sagan. "Health Effects of Energy Production and Conversion," *Annual Review of Energy* 1, 581 (1976).

57. M.G. Morgan, B.R. Barkovich, and A.K. Meier. "The Social Costs of Producing Electric Power from Coal: A First-Order Calculation, *Proceedings of IEEE* 61, 40 (1973).

58. L.A. Sagan. "Health Costs Associated with the Mining, Transport, and Combustion of Coal in the Steam-Electric Industry," *Nature* 250, 107 (1974).

59. *Mineral Resources and the Environment*, Section III, Chapters IX and X. Committee on Mineral Resources and the Environment (COMRATE), Commission on Natural Resources, National Research Council, National Academy of Sciences, Washington, D.C. (1975), p. 203ff. (See Reference 37.)

60. *The Economics of Nuclear Power: A New England Perspective.* Energy Policy Office, Commonwealth of Massachusetts (December 1975).

61. Allied General Nuclear Service. "Nuclear Fuel Cycle Closure Alternatives" (April 1976).

62. J. Roberts. ERDA 15th Annual Nuclear Engineering Education Conference, Argonne National Laboratory, March 1976.

63. A.S. Kubo and D.J. Rose. "Disposal of Nuclear Wastes," *Science* 182, 1205 (1973).

64. J.V. Blomeke, J.P. Nichols, and W.C. McClain. "Managing Radioactive Wastes," *Physics Today* 26, (August 1973), p. 36.

65. "Report to the American Physical Society by the Study Group on Light-Water Reactor Safety," *Reviews of Modern Physics* 47, Supplement No. 1 (1975).

66. J.G. Asbury. "Problems in the Commercialization of New Energy Conservation Technologies," presented at the conference Energy Conservation: a National Forum, Dec. 1–3, 1975, Ft. Lauderdale, Florida.

Index

A

Accidents, cost of, 10, 23-25, 43-46, 102, fig. 4-9
Acid drainage, cost of, 19, 97
Air temperature, 20
Alabama Power Company, 35, 36, 38, 46
Algal growth, 20
Argonne National Laboratory, 95, 96
Atomic Energy Commission, 96

B

Boiling water reactors (BWRs), 8
Boston, 2
Breeder reactors, 23, 24, 30, 36, 57, 94
 fuel costs of, 96

C

Canadian deuterium (CANDU) system, 9
Capacity factors
 and pollution, 22
 of power plants, 18
Capital investment costs, 14, 17, 31-42, 77, 79, figs. 4-3 to 4-11; table 4-21
Carbon dioxide, 21, 99
Carrying charges, 17
Chicago, 2
Chicago Commonwealth Edison Co., 4, 34, 36, 38, 46, 57, 79-85, 93, 95, tables 4-21 to 4-26

Clean Air Act Amendments of 1977, 28, 93
Cleveland Electric, 36, 38, 46
Climate changes, cost of, 10, 98-99
Coal fuel, 28. *See also* Fossil fuel
 amounts used, 73, table 4-19
 cleaning, 18
 cost of, 2, 5-6, 95-96
 and property damage, 97
 and risk aversion factors, 76-79
 supplies of, 72, 75
 in Wyoming, 2
Coal plants
 accident costs in, 24
 capital costs of, 31-42, figs. 4-3 to 4-11
 construction costs of, 93-94
 fuel costs of, 46, 49, fig. 4-11
 human costs of, 43-46, fig. 4-9
 and new technology costs, 46, 4-10
Cohen, B., 22, 23
Construction, costs, 17, 93-94
Cooling towers, 17
Corner solutions, 34
Cost(s). *See also* Social costs
 of acid drainage, 19, 97
 of accidents, 10, 23-25, 102
 apparent, 1
 capital, 14, 17, 31-42, 77, 79, figs. 4-3 to 4-11; table 4-21
 of climate changes, 10, 19-21, 98-99
 of construction, 17, 93-94

121

Index

contributing, 14-23
cumulative, fig. 4-2
of decommissioning, 17
and demand, 27-28, 49, tables 4-1 to 4-4
of development, 10, 26
direct, 9-10, 14-18, 22, 82, 84, 93-97, tables 4-24 to 4-26
of diversity, 88
of electricity, 5
external, 14
of fuel, 10, 14, 17-18, 25, 26, 46, 49, 77, 82, 95-96, fig. 4-11; tables 4-1 to 4-4, 4-23
  coal, 5-6, 95-96
  nuclear, 5-6, 18, 95-96
  transporting, 22
of health, 10, 21-23, 27, 99-101
of heat loss, 19-21, 98-99
hidden, 1
human, 57, 68, 77, 99-101, tables 4-2 to 4-4, 4-9
incremental, 49, tables 4-5 to 4-9, 4-14 to 4-18
indirect, 10, 18-21, 97-99
of insurance, 45
internal, 14
of labor, 18n
of land subsidence, 19, 97
mining-related, 19, 22, 97
of operation and maintenance, 14, 81, 94-95, table, 4-22
of pollution, 3, 14, 19-20, 22, 97, 99, 100
of property damage, 3, 10, 19, 25, 97-98
of reclamation, 19, 97
of regulation, 26
of research, 3, 10, 14, 18-19, 57, 60, 97, tables 4-1 to 4-17
of sabotage, 10, 23-25, 102-103
of safety, 10, 14, 21-23, 27, 99-101
of spent fuel storage, 3, 10, 22-23, 26, 100-101
of technology, 3, 18-19, 36, 46, 49, 57, 60, 94, figs. 4-3 and 4-10, tables 4-1 to 4-17
of terrorism, 23, 25
total, 49-50, 57, 68, 87-88, tables 4-1 to 4-4, 4-10 to 4-13
of water, 10
Council on Economic Priorities, 5

D

Decommissioning, costs, 17
Demand, 10
and capital costs, 36, 38, fig. 4-4
and costs, 27-28, 49, tables, 4-1 to 4-4
as exogenous, 5, 8
growth rates of, 1-2, 28, fig. 4-1
and power production, 13
Development, cost of, 10
Discount rates, 10, 13, 27
and capital costs, 38, figs. 4-5 and 4-6
and costs, 49, 57, 77, tables 4-2 to 4-5
and human costs, 45-46
Distribution, costs, 14
Diversion, of nuclear fuel, 2
Duke Power Company, 35, 36, 38, 46, 94

E

Earnings, discounted future, 21
Electricity
  alternative methods of producing, 8
  amounts supplied, table 4-19
  costs, 5
  selecting sources of, 6-7
  and nuclear weapons proliferation, 10
Emissions, harmful, 22
Energy conservation, 8
Environmental Protection Act, 82, 97
Eutrophication, 20

F

Federal Power Commission National Power Survey, 28
Federal Power Commission Report (1975), 35
Fishing yields, 20
Ford-Mitre study, 5, 6
Fossil fuel, and health hazards, 22. *See also* Coal fuel
Fuel. *See also* Coal fuel; Nuclear fuel; Oil fuel; Uranium 2
  costs of, 1-2, 10, 14, 17-18, 46, 49, 77, 82, 95-96, fig. 4-11; tables 4-1 to 4-4, 4-23
  storing spent, 10, 22-23, 100-101
  supplies, 72
  transportation costs of, 22

G

Generating stations. *See* Power plants
Greenhouse effect, 99
Growth rates
  and capital costs, 38, fig. 4-7
  of electricity demand, 28, fig. 4-1

H

Health. *See also* Human costs
  costs of, 10, 21-23, 43-46, fig. 4-9
  and discount rates, 27
Heat islands, 20
Heat loss, costs of, 19-21, 98-99
Heavy water reacrors (HWRs), 9
Human costs, 57, 68, 77, 99-102, tables 4-2 to 4-4, 4-9
Human life
  discount rates for, 13
  valuation of, 21-22
Hydroelectric power, 24, 28

I

Insurance, costs of, 45, 102
Isocost curves, 31

L

Labor, costs and coal, 18n
Land subsidence, costs, 19, 97
Light water reactors (LWRs), 9, 23, 24, 28, 29, 72, 75, 93
Lifetimes, of power plants, 28

M

Maintenance, costs of, 14, 81, 94-95, table 4-22
Manne, A., 5, 57
Mining, costs of, 19, 22, 97
Mishan, E.J., 21

N

National Center for Atmospheric Research, 20
New York, 2
Nuclear Energy Policy Study Group report, 5
Nuclear fuel. *See also* Uranium
  costs of, 5-6, 18
  diversion of, 2
  in Northeast, 2
  resistance to, 24-25
  and risk aversion factors, 76-79
  safety, 18
  storage costs of spent, 3, 10, 22-23, 26, 100-101
  and waste disposal, 18
Nuclear plants
  accident costs in, 23-25
  capital costs of, 31-42, figs. 4-3 to 4-11
  construction costs of, 93-94
  fuel costs of, 46, 49, fig. 4-11
  human costs of, 43-46, fig. 4-9
  and new technology costs, 46, fig. 4-10
  security against sabotage costs in, 23-25
Nuclear Regulatory Commission (NRC), 26
Nuclear weapons, proliferation of, 8, 10

O

Operation, costs of, 14, 22, 81, 94-95, table 4-22
Optimum mix
  variables affecting, 1-4, 9-10, 87
  and human costs, 57
  and sensitivity to discount rates, 27
Oil fuel, 28

P

Pay, willingness to, 21
Philadelphia Electric, 93, 94
Pigford, T.H., 72-73
Pigou, A.C., 6-7
Plutonium, 72, 96
Policy making, using TOSCA for, 2
Pollution, costs of, 3, 14, 19, 22, 97
Power plants. *See also* Coal plants; Nuclear plants
  capacity factors of, 18
  construction of, 30-31
  decommissioning costs of, 17
  lifetimes of, 17, 28
Pressurized-water reactors (PWRs), 8
Price
  and demand, 5
  estimating future, 18
Price-Anderson Act, 45, 102
Power production, demand, 13
Project Independence Report, 28
Proliferation, of nuclear weapons, 10
Property damage, costs of, 3, 10, 19, 97-98
Property taxes, 17

R

Rasmussen Report, 43, 45, 102
Reactors
  boiling water, 8
  breeder, 23, 24, 30, 36, 57, 94
  light water, 9, 23, 24, 29, 72, 75, 93
  oil-fired, 24
Reclamation, costs of, 19, 97
Regulation, costs of, 26
Region, and optimum mix, 1-2
Repairs, 17

Research, costs of, 3, 10, 14, 18, 57, 60, 97, tables 4-1 to 4-17. *See also* Technology

Risk
 aversion factors, 23, 24, 25, 45, 76-79
 payment for, 21

S

Sabotage, costs of safeguarding against, 1, 10, 102-103
Safeguards, social effects of, 10-11
Safety. *See also* Human costs
 costs of, 10, 14, 21-23, 43-46, 99-101, fig. 4-9
 and discount rates, 27
 and nuclear fuel, 18
Schelling, T.C., 21
Scrubbers, 28, 82
Searl, M.F., 72
Social benefits, estimating, 6
Social costs, 6, 14, 27. *See also* Costs
Solar constant, changes in, 20-21
Solar power, 30, 36, 94
 maintenance costs of, 95
 fuel costs of, 96
Storage, spent nuclear fuel costs of, 3, 10, 22-23, 26, 100-101
Sulfur dioxide, 22

T

Taxes, property, 17
Technology. *See also* Research
 and capital costs, 36, 46, figs. 4-3, 4-10
 and coal plant construction, 30-31
 costs of, 3, 18-19, 49, 57, 60, 94, tables 4-1 to 4-17
 and nuclear plant construction, 30-31
Temperature
 of air, 20
 changes in, 20-21, 98-99
Terrorism, costs of, 23, 25
Total Social Cost Analysis (TOSCA), using, 2, 84, 88
Total social costs, 49-50, 57, 68, 87-88, tables 4-1 to 4-4, 4-10 to 4-13

U

Uranium, 28, 29. *See also* Nuclear fuel
 amount used, 73-74, tables 4-19 and 4-20
 cost of, 2, 49, 95-96
 miners of, 43, 45
 supplies, 72, 75

W

Wages, 17
Water, cost of, 10
Wyoming, 2

Y

Yang, C.S., 72-73

About the Authors

Linda L. Gaines is a physicist who received her graduate degrees from Columbia University. Her interest turned to energy-related matters when she began her post-doctoral research at the University of Chicago. Presently, she is examining energy implications of material substitution in the Energy and Environmental Systems Division of Argonne National Laboratory.

R. Stephen Berry is a physical chemist with fourteen years experience in resource policy. He is author or coauthor of over one hundred scientific papers, including many on energy, environment and other resources. He is a member of the Board of Directors of the Bulletin of the Atomic Scientists, an Advisory Editor of *Resources and Energy* and a consultant or advisor to Argonne National Laboratory, Los Alamos Scientific Laboratory and Oak Ridge National Laboratory. At the University of Chicago he is Professor in the Department of Chemistry, the James Franck Institute the Committee on Public Policy Studies.

Thomas Veach Long, II has held faculty appointments at The Medical College of Virginia (biophysics), The Pennsylvania State University (chemistry), and The University of Michigan (chemistry). While at The University of Michigan, he was also a Visiting Scholar in Economics. Currently, he is a member of the faculty of The Committee on Public Policy Studies at The University of Chicago, holding the position Research Associate (Associate Professor). His research interests lie in the areas of science and technology policy, energy analysis, and resource economics, and in the integration of physical information into economic behavioral relationships. He is the author of more than 45 scientific and economic papers, and is co-editor of the new interdisciplinary journal, *Resources and Energy*.

THE LIBRARY
ST. MARY'S COLLEGE OF MARYLAND
ST. MARY'S CITY, MARYLAND 20686